Quantum Theory of Solids

T0176053

The Taylor & Francis *Masters Series in Physics and Astronomy*

Edited by David S. Betts
Department of Physics and Astronomy, University of Sussex, Brighton, UK

Core Electrodynamics
Sandra C. Chapman
0-7484-0622-0 (PB)
0-7484-0623-9 (HB)

Atomic and Molecular Clusters
Roy L. Johnston
0-7484-0931-9 (PB)
0-7484-0930-0 (HB)

Quantum Theory of Solids
Eoin P. O'Reilly
0-7484-06271-1 (PB)
0-7484-0628-X (HB)

Forthcoming titles in the series:

Basic Superfluids
Anthony M. Guénault
0-7484-0892-4 (PB)
0-7484-0891-6 (HB)

Quantum Theory of Solids

Eoin P. O'Reilly

CRC Press
Taylor & Francis Group
Boca Raton London New York

CRC Press is an imprint of the
Taylor & Francis Group, an **informa** business

First published 2002 by Taylor & Francis

Published in 2023 by CRC Press
Taylor & Francis Group
6000 Broken Sound Parkway NW, Suite 300
Boca Raton, FL 33487-2742

ISBN 13: 978-0-7484-0627-2 (pbk)

Visit the Taylor & Francis Web site at
http://www.taylorandfrancis.com

and the CRC Press Web site at
http://www.crcpress.com

Typeset in 10/12 pt Palatino by
Newgen Imaging Systems (P) Ltd, Chennai, India

British Library Cataloguing in Publication Data
A catalogue record for this book is available
from the British Library

Library of Congress Cataloging in Publication Data
O'Reilly, Eoin P.
  Quantum theory of solids/Eoin P. O'Reilly.
    p.cm. – (Masters series in physics and astronomy)
  Includes bibliographical references and index.
  ISBN 0-7484-0628-X (alk. paper) – ISBN 0-7484-0627-1 (pbk.: alk. paper)
  1. Solid state physics. 2. Quantum theory. 3. Chemical bonds. I. Title. II. Series.

QC176. O74 2002
530.4'—dc21   200205724

ISBN 0-7484-0628-X

Contents

Series preface

The Masters series of textbooks is aimed squarely at students taking specialised options in topics within the primary areas of physics and astronomy, or closely related areas such as physical chemistry and environmental science. Appropriate applied subjects are also included. The student interest group will typically be studying in the final year of their first degree or in the first year of postgraduate work. Some of the books may also be useful to professional researchers finding their way into new research areas, and all are written with a clear brief to assume that the reader has already acquired a working knowledge of basic core physics.

The series is designed for use worldwide in the knowledge that wherever physics is taught at degree level, there are core courses designed for all students in the early years followed by specialised options for those consciously aiming at a more advanced understanding of some topics in preparation for a scientific career. In the UK there is an extra year for the latter category, leading to an MPhys or MSc degree before entry to postgraduate MSc or PhD degrees, whereas in the USA specialisation is often found mainly in masters or doctorate programmes. Elsewhere the precise modulations vary but the development from core to specialisation is normally part of the overall design.

Authors for the series have usually been able to draw on their own lecture materials and experience of teaching in preparing draft chapters. It is naturally a feature of specialist courses that they are likely to be given by lecturers whose research interests relate to them, so readers can feel that they are gaining from both teaching and research experience.

Each book is self-contained beyond an assumed background to be found in appropriate sections of available core textbooks on physics and useful mathematics. There are of course many possibilities, but examples might well include Richard P. Feynman's three-volume classic *Lectures on Physics* (first published by Addison-Wesley in the 1960s) and Mary L. Boas' *Mathematical Methods in the Physical Sciences* (Wiley, 1983). The primary aim of books in this series will be to enhance the student's knowledge base so that they can approach the research literature in their chosen field with confidence. They are not intended as major treatises at the forefront of knowledge, accessible only to a few world experts; instead they are student-oriented, didactic in character and written to build up the confidence of their readers. Most volumes are quite slim and they are generously illustrated.

Different topics may have different styles of questions and answers, but authors are encouraged to include questions at the end of most chapters, with answers at the end of the book. I am inclined to the view that simple numerical answers, though essential, are often too brief to be fully satisfactory, particularly at the level of this series. At the other extreme, model answers of the kind that examination boards normally require of lecturers would make it too easy for a lazy reader to think they had understood without actually trying. So the series style is to include advice about difficult steps in calculations, lines of detailed argument in cases where the author feels that readers may become stuck, and any algebraic manipulation which might get in the way of proceeding from good understanding to the required answer. Broadly, what is given is enough help to allow the typical reader to experience the feelgood factor of actually finishing questions, but not so much that all that is needed is passive reading of a model answer.

David S. Betts
University of Sussex

Preface

The application of quantum theory to solids has revolutionised our understanding of materials and their applications. This understanding continues to drive the development of the functional materials which form the basis of modern technology. This book aims to describe the physics of the electronic structure of these materials.

There are already many excellent texts that provide a first introduction to solid state physics, and others that develop a more advanced understanding of the subject. Why, then, another text? This book is based on final year undergraduate and first year postgraduate lectures that I have presented over the last twelve years, originally in the Department of Physics at the University of Surrey and now in University College Cork. My motivation for the book was based primarily on there being no one text that I found suitable for the lecture courses I was presenting. The lecture courses aimed to provide a self-contained description that focuses on electronic structure, and addresses three of the most important topics in solid state physics: semiconductors, magnetism and superconductivity.

As an advanced undergraduate text, this book assumes pre-knowledge in several of the main areas of physics, including quantum mechanics, electromagnetism, thermal physics and an introductory course in the properties of matter. However, the first time I gave the course, I assumed not just that the students had covered such material in lectures but also that they were still familiar with it, in many cases over a year after the previous lecture courses had finished. This was a mistake, and convinced me in subsequent years to begin by reviewing quantum concepts relevant to an undergraduate solid state physics course, going from the basics through to a relatively advanced level, and including techniques such as the variational method and perturbation theory.

This initial revision of quantum mechanics provides some of the key foundations for the remainder of the book. The variational method justifies many of the approximations that we use to describe electronic structure, and to understand the electronic band structure and chemical bonding trends in isolated molecules and in crystalline solids. As examples, the main trends in molecular bonding are developed by considering a double square well. The nearly-free-electron and tight-binding methods are both introduced by using the Kronig–Penney model with an infinite linear chain of square wells, which is then also used to explain the concept of a pseudopotential.

The material in the book extends in several places topics covered in the lecture courses at Surrey and Cork. Material from the first five chapters was used originally as a 30-hour undergraduate and then a 20-hour postgraduate introduction to the electronic structure and applications of advanced semiconductor materials. Selected material from Chapters 1, 2 and 5 was also combined with the last three chapters to present a self-contained 20-hour final year undergraduate course on the "Quantum Theory of Solids."

The content of this book is more limited than others with this title. With the focus on functional materials, less emphasis is placed on the electronic properties of metals. There is also little consideration of the vibrational and dynamical properties of solids, nor of their dielectric response. These were all omitted to keep the book to a reasonable length (and cost). Finally, although Bloch's theorem and the wavevector k underpin much of the analysis, less emphasis is placed on the concept of reciprocal lattice and its use for determining structural properties through diffraction studies. This omission was deliberate. I have found that it is difficult to visualise how a reciprocal lattice relates to its real-space counterpart, particularly in three dimensions; this difficulty can then distract from understanding many trends in the electronic structure of solids. When possible, the reciprocal lattice is, therefore, used predominantly in one- and two-dimensional examples, where it is generally more straightforward to picture how the real-space and reciprocal lattices relate to each other.

I am very grateful to all those who have helped in the preparation and production of this book. These include many students at Surrey, in particular Martin Parmenter, Andy Lindsay and Gareth Jones, who worked through and helped to develop several of the problems. I thank Joy Watson for her help with copyright requests, Patricia Hegarty for help in producing the index, and Vincent Antony at Newgen for his efficient and helpful handling of the proofs. I thank Dave Faux, Betty Johnson, Fedir Vasko and David Betts for their careful reading and comments on the complete text, and James Annett, Dermot Coffey and Maurice Rice for their feedback and comments on the chapter on superconductivity. Much of the material on electronic structure and semiconductors developed from extended discussions and interactions with Alf Adams. Last but not least I thank Anne and my family for their support and encouragement while I was writing this book.

Chapter 1

Introduction and review of quantum mechanics

1.1 Introduction

The application of quantum theory to solids has revolutionised our understanding of materials and played a pivotal role in the information revolution of the last fifty years. At the most basic level, quantum theory enables us to understand why some solids are metals, some insulators, and some semiconductors. It also allows us to understand trends in the properties of different materials and to engineer materials to the properties we desire.

The exact features of the electronic band structure of semiconductor materials, for instance, play a key role in determining their electronic and optoelectronic properties, and their usefulness for devices such as lasers, detectors, and electronic integrated circuits. Details of the exchange interaction between electrons on neighbouring atoms determine the differences between ferromagnets, antiferromagnets and ferrimagnets and their applicability for data storage or transformer applications. Quantum mechanical properties are measurable on a macroscopic scale in superconductors, allowing both the determination of fundamental constants and the commercial development of technologies such as magnetic resonance imaging.

The understanding and development of the functional solids which form the basis of modern technology has been achieved through a synergy between physicists, material scientists, and engineers. As the title implies, this book is primarily concerned with the physicist's story: we first review the basic concepts of quantum mechanics and quantum mechanical techniques, in order to use them later to understand some of the wide variety of electronic properties of solids and their applications.

The remainder of this chapter is concerned with a review of quantum mechanics: much of the material may be familiar from previous courses and can certainly be found in many other textbooks, some of which are listed at the end of this chapter. Providing an overview here should at least refresh some of the key concepts in quantum mechanics. It may also introduce techniques and examples which have not previously been studied, and which are particularly useful towards understanding trends in the quantum theory of solids.

1.2 Wave–particle duality

The idea that light could be described either as a wave or as a particle was formally introduced by Einstein in 1906. He deduced from the photoelectric effect that light waves also have particle-like properties, with the energy E of an individual light packet, or photon, given by

$$E = h\nu \tag{1.1}$$

where h is Planck's constant and ν is the frequency of the light. It is relatively easy to demonstrate experimentally that light has both wave-like and particle-like properties: the diffraction patterns seen with Young's slits or in a Michelson interferometer are characteristic of wave interference, while effects such as Compton scattering or the existence of an absorption edge in a semiconductor are best explained by regarding a light beam as made up of individual packets, photons, each of which has particle-like properties.

It is less obvious that objects which we regard as particles can also have wave-like properties. It took nearly twenty years after the discovery of the photoelectric effect before de Broglie postulated in 1924 that particles can also have wave-like properties. He associated a wavelength λ with a particle of momentum p, with the two being related through Planck's constant:

$$\lambda = h/p \tag{1.2}$$

If we consider a macroscopic object of mass 1 kg moving with unit velocity, $1\,\mathrm{m\,s^{-1}}$, then its de Broglie wavelength, 6×10^{-34} m is clearly of a length scale which will not be readily detectable, and so the object's motion can be treated classically, without regard for the wave-like properties. If, however, we consider an electron with mass m of order 10^{-30} kg, whose kinetic energy $p^2/2m$ is comparable to the room temperature thermal energy, $kT(\approx 25\,\mathrm{meV})$, then the de Broglie wavelength, $\lambda = h/(2mkT)^{1/2}$, is of order $12\,\text{Å}$, comparable to the typical interatomic spacing in a solid ($\approx 3\,\text{Å}$). We might, therefore, expect and indeed do find wave-like properties, including reflection and diffraction, playing a large role in determining the behaviour of electrons in solids.

What is the property of matter which behaves as a wave? We call the quantity, whose variation makes up matter waves, *the wavefunction*, symbolised as $\Psi(x, y, z, t)$, giving the amplitude of the wave at a given point in space and time. It is reasonable to expect the wavefunction Ψ to be related

Figure 1.1 The time-dependent variation in the amplitude of a cosine wave at a fixed position in space ($x = 0$).

to the probability P of finding the particle at (x, y, z) at time t. As Ψ is the wave amplitude and amplitudes can be greater or less than zero, while a probability P always lies between 0 and 1, we cannot associate Ψ directly with P. Born postulated in 1926 that the probability of finding the particle at (x, y, z) at time t is proportional to the value of $|\Psi(x, y, z, t)|^2$, as $|\Psi|^2 \geq 0$. The wavefunction is therefore sometimes more properly referred to as the probability amplitude Ψ, and may indeed be described by a complex function, as $|\Psi|^2 = \Psi^*\Psi \geq 0$ for complex numbers.

As quantum mechanics treats particles as waves, it is also useful at this point to review some of the key properties of waves.

The simplest equation for the variation in the amplitude y of a wave as a function of time t is

$$y(t) = A\cos(2\pi \nu t) \tag{1.3}$$

which is illustrated in fig. 1.1, where ν is the frequency of the wave and A its maximum amplitude. The period T over which the wave repeats itself is equal to $1/\nu$. Such a wave may also be written as

$$y(t) = A\cos(\omega t) \tag{1.4}$$

where ω is the angular frequency. When we talk about a wave propagating along the x-axis with velocity u, we mean that a particular crest moves a distance $x = ut$ in time t, so that the displacement y at position x and time t is then the same as the displacement at $x = 0$ at time $t - x/u$, which from eq. (1.3) is given by

$$y(x, t) = A\cos 2\pi \nu(t - x/u)$$
$$= A\cos 2\pi(\nu t - \nu x/u) \tag{1.5}$$

As the velocity $u = \nu\lambda$, the displacement in eq. (1.5) is also given by

$$y(x, t) = A\cos(2\pi \nu t - 2\pi x/\lambda) \tag{1.6}$$

which can be written in more compact form by introducing the wave-number $k = 2\pi/\lambda$, to give

$$y(x, t) = A\cos(\omega t - kx) \tag{1.7}$$

This can be generalised to a plane wave propagating along the direction n in three dimensions as

$$y(r,t) = A\cos(\omega t - k \cdot r) \tag{1.8}$$

where the magnitude of the wavevector k is determined by the wavelength, $|k| = 2\pi/\lambda$, while the direction along which k is pointing equals the propagation direction n.

1.3 Wave quantisation

While a wave propagating in free space may in principle have any wavelength, once it is confined within a finite region it must satisfy definite boundary conditions, so that only certain modes are supported, as is indeed found for vibrations on a string or in musical instruments. If we consider a particle of mass m confined by an infinite potential between $x = 0$ and $x = L$, we expect its amplitude to go to zero at the boundaries, as for a vibration on a string. The allowed wavelengths λ_n are then given by

$$\lambda_n = \frac{2L}{n} \tag{1.9}$$

where n is a positive integer, $n = 1,2,3,\ldots$. The first three allowed states are illustrated in fig. 1.2. But, de Broglie postulated that the momentum of a particle with wavelength λ is given by $p = h/\lambda$, so we find for the nth allowed state that

$$p_n = \frac{hn}{2L} \tag{1.10}$$

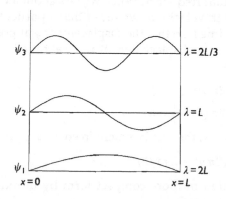

Figure 1.2 The first three allowed standing waves for a particle in a box of width L (or vibrational modes of a string of length L).

and the energy is then quantised, with only certain discrete energies allowed:

$$E_n = \frac{p_n^2}{2m} = \frac{h^2 n^2}{8mL^2}$$

(1.11)

It is because of this quantisation of wavelength and energy that the wave-like description of matter is generally referred to as 'wave mechanics' or 'quantum mechanics'.

1.4 Heisenberg uncertainty principle

Once we adopt a wave-like description of matter, it is then no longer possible to describe a particle as being just at one particular point in space. A particle must now be described by a wavepacket, and any wavepacket is always spread over some region of space. Figure 1.3 illustrates two wave packets with similar wavelength (and therefore similar average momentum): one wavepacket is confined within a small region Δx of space, of order one wavelength long, while the second wavepacket is defined over a much larger region, and therefore has a far greater uncertainty, Δx, in its position. Each of these packets can be defined as a sum (strictly an integral) over plane waves with different wavevectors, k. To achieve a tightly defined wavepacket, such as the upper wave in fig. 1.3, it is necessary to include waves with a wide range of wavevectors k; the range of wavevectors in the upper case, Δk, is then much larger than for the lower wavepacket. It can in fact be proved for any wavepacket that

$$\Delta x \, \Delta k \geq \tfrac{1}{2}$$

(1.12)

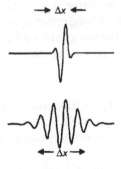

Figure 1.3 Two wavepackets of comparable wavelength, but different spatial extent, Δx. We need to include many components with different wavevectors, k, in the upper, compact, wavepacket, so that Δk is large for small Δx. By contrast, the lower wavepacket extends over several wavelengths, so the range of wavevectors, Δk, is small in this case for large Δx.

Replacing λ in eq. (1.2) by the wavenumber k, we find that the momentum, p, of a wave is directly related to its wavenumber by

$$p = hk/2\pi = \hbar k \tag{1.13}$$

where we introduce $\hbar = h/2\pi$. Substituting eq. (1.13) in (1.12), we derive Heisenberg's uncertainty principle, one of the most widely quoted results in quantum mechanics:

$$\Delta x \, \Delta p \geq \hbar/2 \tag{1.14}$$

namely that it is impossible to know the exact position and exact momentum of an object at the same time.

A similar expression can be found relating energy E and time t. We saw that the energy E is related to a frequency ν by $E = h\nu$. The uncertainty in measuring a frequency ν depends on the time Δt over which the measurement is made

$$\Delta \nu \sim 1/\Delta t \tag{1.15}$$

so that the uncertainty in energy $\Delta E = h\Delta \nu \sim h/\Delta t$, which can be re-arranged to suggest $\Delta E \Delta t \sim h$. When the derivation is carried out more rigorously, we recover a result similar to that linking momentum and position, namely

$$\Delta E \Delta t \geq \hbar/2 \tag{1.16}$$

so that it is also impossible to determine the energy exactly at a given moment of time.

1.5 Schrödinger's equation

Although it is impossible to *derive* the equation which determines the form of the wavefunction Ψ, Schrödinger was nevertheless able to deduce or postulate its form. We discussed above how $\Psi(x,t)$ may be given by a complex function. We assume we can choose

$$\Psi(x,t) = A\,e^{-i(\omega t - kx)} \tag{1.17}$$

for a wave propagating in the x-direction with angular frequency ω and wavenumber k. Using eqs (1.1) and (1.2), we can rewrite ω and k, and hence Ψ in terms of the energy E and momentum p, respectively:

$$\Psi(x,t) = A\,e^{-i/\hbar(Et - px)} \tag{1.18}$$

We can then take the partial derivative of the wavefunction with respect to time, t, and find

$$\frac{\partial \Psi}{\partial t} = -\frac{iE}{\hbar}\Psi$$

which can be re-arranged to give

$$E\Psi = i\hbar \frac{\partial \Psi}{\partial t}$$

(1.19)

while taking the second derivative with respect to position x, we find

$$\frac{\partial^2 \Psi}{\partial x^2} = -\frac{p^2}{\hbar^2}\Psi$$

or

$$p^2\Psi = -\hbar^2 \frac{\partial^2 \Psi}{\partial x^2}$$

(1.20)

Classically, the total energy E of a particle at x is just found by adding the kinetic energy $T = p^2/2m$ and potential energy V at x:

$$E = \frac{p^2}{2m} + V$$

(1.21)

Schrödinger assumed that if you multiply both sides of eq. (1.21) by the wavefunction Ψ, the equation still holds:

$$E\Psi = \left(\frac{p^2}{2m} + V\right)\Psi$$

(1.22)

Then substituting eqs (1.19) and (1.20) into (1.22), Schrödinger postulated that the wavefunction Ψ obeys the second order partial differential equation

$$i\hbar \frac{\partial \Psi}{\partial t} = -\frac{\hbar^2}{2m}\frac{\partial^2 \Psi}{\partial x^2} + V\Psi$$

(1.23)

This is referred to as Schrödinger's time-dependent (wave) equation; the 'proof' of its validity comes from the wide range of experimental results which it has predicted and interpreted.

For many problems of interest, the potential $V(x)$ does not vary with time and so we can separate out the position- and time-dependent parts of Ψ:

$$\Psi(x, t) = \psi(x)\, e^{-iEt/\hbar}$$

(1.24)

Substituting eq. (1.24) in (1.23), and then dividing through by $e^{-iEt/\hbar}$ gives

$$-\frac{\hbar^2}{2m}\frac{d^2 \psi(x)}{dx^2} + V(x)\,\psi(x) = E\psi(x)$$

(1.25a)

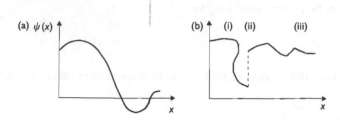

Figure 1.4 (a) A 'well-behaved' (=allowed) wavefunction is single-valued and smooth (i.e. ψ and $d\psi/dx$ continuous). (b) This is certainly *not* a wavefunction, as it is (i) multiple-valued, (ii) discontinuous, and (iii) has discontinuous derivatives.

often rewritten as

$$H\psi(x) = E\psi(x) \qquad (1.25b)$$

where $H = (-\hbar^2/2m)d^2/dx^2 + V(x)$ is referred to as the Hamiltonian operator. (The name arises because it is a function which acts, or 'operates' on ψ.) Equation (1.25) is referred to as the time-independent, or steady-state Schrödinger equation. It is a second order ordinary differential equation. We expect solutions of such an equation to generally behave 'sensibly', which can be expressed mathematically, and is illustrated in fig. 1.4, as requiring that

1   ψ is a single-valued function (otherwise, there would be a choice of values for the probability function $|\psi(x)|^2$);
2   ψ is continuous; and
3   $d\psi/dx$ is continuous.

Similar conditions generally apply to the solutions of other second order differential equations, such as that describing the vibrational modes of a fixed string. As with the vibrational modes of a fixed string, which occur only at certain well-defined frequencies, Schrödinger's equation only has allowed solutions for certain well-defined energies E in a potential well, $V(x)$.

There are remarkably few potentials $V(x)$ in which Schrödinger's equation is analytically soluble: those which exist are therefore very useful and instructive to study. Examples where analytical solutions can be found include free space, the infinite square well and finite square well, the hydrogen atom, and the simple harmonic oscillator. All of these, and related examples, are used later to elucidate aspects of the quantum theory of solids.

1.6 Expectation values and the momentum operator

Schrödinger's equation can in principle be solved for an arbitrary potential, V, giving a set of allowed energy levels E_n with associated wavefunctions, ψ_n. As we wish to associate $|\psi_n(x)|^2$ with the probability distribution of the particle, and the particle has a 100 per cent chance (probability $= 1$) of being somewhere along the x-axis, it is customary to 'normalise' the wavefunction $\psi_n(x)$ so that

$$\int_{-\infty}^{\infty} |\psi_n(x)|^2 \, dx = 1 \tag{1.26}$$

and the probability of finding a particle in the nth state between x and $x + dx$ is then given by

$$P_n(x)dx = |\psi_n(x)|^2 \, dx \tag{1.27}$$

as illustrated in fig. 1.5. The *expectation* (or average) value $\langle x_n \rangle$ of the position x for a particle with wavefunction $\psi_n(x)$, is then found by evaluating the integral

$$\langle x_n \rangle = \int_{-\infty}^{\infty} x |\psi_n(x)|^2 \, dx \tag{1.28a}$$

which can also be written as

$$\langle x_n \rangle = \int_{-\infty}^{\infty} \psi_n^*(x) x \psi_n(x) \, dx \tag{1.28b}$$

Although both forms of eq. (1.28) give the same result, it can be shown that the second form is the correct expression to use. The expectation value for an arbitrary function $G(x)$ is then given by

$$\langle G_n \rangle = \int_{-\infty}^{\infty} \psi_n^*(x) G(x) \psi_n(x) \, dx \tag{1.29}$$

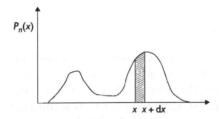

Figure 1.5 Plot of the probability distribution function, $P_n(x) = |\psi_n(x)|^2$ for a normalised wavefunction, $\psi_n(x)$. The total area under the curve, $\int_{-\infty}^{\infty} P_n(x)dx$, equals 1, and the probability of finding the particle between x and $x + dx$ is equal to the area of the shaded region.

The average momentum, $\langle p \rangle$, can be calculated in a similar way, if we identify the operator, $-i\hbar \partial / \partial x$ with the momentum p. This is suggested by the wavefunction $\Psi(x,t) = Ae^{-i(Et-px)/\hbar}$ for a particle in free space. Taking the partial derivative of this wavefunction with respect to position, x, gives

$$-i\hbar \frac{\partial \Psi}{\partial x} = p\Psi \tag{1.30}$$

The expectation value of the momentum, $\langle p_n \rangle$, is then found by calculating

$$\langle p_n \rangle = \int_{-\infty}^{\infty} \psi_n^*(x) \left(-i\hbar \frac{\partial}{\partial x} \right) \psi_n(x) \, dx \tag{1.31}$$

This expression is used later, in describing the $k \cdot p$ technique for calculating semiconductor band structure, and also when considering current flow in superconductors.

1.7 Some properties of wavefunctions

There are several 'tricks' based on the general properties of wavefunctions which are useful when solving and applying Schrödinger's equation. It is not intended to derive these properties here, but it is nevertheless useful to review them for later applications, and to provide the weakest type of 'proof' for each of the properties, namely proof by example.

Even and odd symmetry: If we have a mirror plane in the potential, chosen at $x = 0$, such that $V(x) = V(-x)$ in one dimension, or $V(x,y,z) = V(-x,y,z)$ in three dimensions, then all of the wavefunctions $\psi_n(x)$ which are solutions of Schrödinger's equation can be chosen to be either symmetric, that is, even, with $\psi(x) = \psi(-x)$, or antisymmetric, that is odd, with $\psi(x) = -\psi(-x)$, as illustrated in fig. 1.6.

In such a symmetric potential, there should be equal probability of finding a particle at x, or at $-x$, which requires that

$$|\psi(x)|^2 = |\psi(-x)|^2 \tag{1.32}$$

This requirement is clearly obeyed when the wavefunctions are even, with $\psi(x) = \psi(-x)$, or odd, with $\psi(x) = -\psi(-x)$. It can also be shown that the wavefunction for the ground state is always an even function in a symmetric potential. We shall see later that symmetry also simplifies the form of the wavefunctions in periodic solids.

Completeness of wavefunctions: The wavefunctions $\psi_n(x)$ which are solutions to Schrödinger's equation for a particular potential, $V(x)$, form a complete set, that is, any other well-behaved function $f(x)$ defined in the same region of space can be written as a linear combination of the

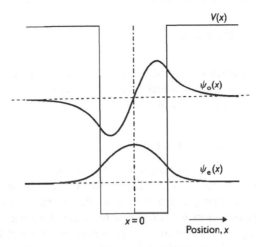

Figure 1.6 A wavefunction, $\psi_e(x)$ of even symmetry and a wavefunction, $\psi_o(x)$ of odd symmetry, in a symmetric potential, $(V(x) = V(-x))$.

wavefunctions $\psi_n(x)$:

$$f(x) = \sum_{n=1}^{\infty} a_n \psi_n(x) \tag{1.33}$$

This is easily seen for the infinite square well potential, whose lowest energy wavefunctions are illustrated in fig. 1.2. There is a standard proof in Fourier analysis that any well-behaved function $f(x)$ defined between 0 and L can be expressed in terms of the Fourier sine series

$$f(x) = \sum_{n=1}^{\infty} a_n \sin \frac{n\pi x}{L} \tag{1.34}$$

But the nth wavefunction in an infinite square well is just given by $\psi_n(x) = (2/L)^{1/2} \sin(n\pi x/L)$, and so the Fourier series proof immediately implies that the infinite well wavefunctions $\psi_n(x)$ also form a complete set.

Orthogonality of wavefunctions: Two energy states with wavefunctions ψ_m and ψ_n are said to be *degenerate* if they have the same energy, $E_m = E_n$. If two states are not degenerate, then it can be shown that their averaged 'overlap', defined by the product $\psi_m^*(x)\psi_n(x)$, is always zero:

$$\int_{-\infty}^{\infty} \psi_m^*(x)\psi_n(x)\, \mathrm{d}x = 0 \tag{1.35}$$

That is, the two wavefunctions are said to be orthogonal to each other. Given a set of degenerate states, it is also always possible to choose wavefunctions for the degenerate states which are orthogonal to each other, so that for a complete set of normalised wavefunctions, we can always write

$$\int_{-\infty}^{\infty} \psi_m^*(x)\psi_n(x)\,\mathrm{d}x = \delta_{mn} \tag{1.36}$$

where δ_{mn}, the Kronecker delta, equals 1 if $m = n$, and is zero otherwise. This result is again readily demonstrated for the wavefunctions $(2/L)^{1/2}\sin(n\pi x/L)$ in an infinite well.

1.8 The variational principle

As the Schrödinger equation cannot be solved analytically for most potentials, it is useful if we can develop techniques which allow straightforward estimates of material properties. The variational method is a particularly important approximation technique, which can be used to estimate the *ground state energy* of a Hamiltonian H where we do not know the exact wavefunctions, $\psi_n(x)$. It can also be used to estimate the energy of the first excited state in a symmetric potential.

For a given arbitrary potential, $V(x)$, it is generally possible to make a reasonable guess, say $f(x)$, for the overall shape and functional form of the ground state wavefunction, $\psi_1(x)$, knowing that the amplitude should be largest near the potential minimum, decaying away to zero as the potential increases (see fig. 1.7). In practice, it is most unlikely that $f(x)$ will be an exact guess for $\psi_1(x)$. But, because of the completeness of the exact wavefunctions, $f(x)$ can always be expressed as in eq. (1.33) in terms of the exact wavefunctions, and the estimated expectation value of the ground state energy, $\langle E \rangle$, can be calculated as

$$\langle E \rangle = \frac{\int_{-\infty}^{\infty} f^*(x)Hf(x)\,\mathrm{d}x}{\int_{-\infty}^{\infty} f^*(x)f(x)\,\mathrm{d}x} \tag{1.37}$$

which is the generalisation of eq. (1.29) for a function $f(x)$ which has not been normalised.

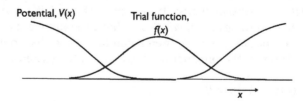

Potential, $V(x)$ Trial function, $f(x)$

x

Figure 1.7 The trial wave function $f(x)$ should be a reasonable guess at the estimated shape of the ground state wavefunction in the arbitrary potential $V(x)$, peaking about the minimum of $V(x)$ and decaying to zero at large x.

The numerator in eq. (1.37) can be expanded using eq. (1.33) in terms of the wavefunctions $\psi_n(x)$, as

$$\int_{-\infty}^{\infty} f^*(x)Hf(x)\,\mathrm{d}x = \int_{-\infty}^{\infty} \left(\sum_m a_m^* \psi_m^*(x)\right)\left(H\sum_n a_n \psi_n(x)\right)\mathrm{d}x$$

$$= \int_{-\infty}^{\infty} \sum_n \sum_m a_m^* a_n \psi_m^*(x)E_n\psi_n(x)\,\mathrm{d}x \qquad (1.38)$$

where we use $H\psi_n = E_n\psi_n$. Using eq. (1.36) for the orthonormality of the wavefunctions, this can be further simplified, giving

$$\int_{-\infty}^{\infty} f^*(x)Hf(x)\,\mathrm{d}x = \sum_{m,n} a_m^* a_n E_n \delta_{mn} = \sum_n |a_n|^2 E_n$$

$$\geq E_1 \sum_n |a_n|^2 \qquad (1.39)$$

as the ground state is by definition the lowest energy state, so that $E_n \geq E_1$ for all values of $n \geq 1$. Using the orthogonality condition, eq. (1.36), it can be readily shown that the denominator of eq. (1.37) is given by

$$\int_{-\infty}^{\infty} f^*(x)f(x)\,\mathrm{d}x = \sum_n |a_n|^2 \qquad (1.40)$$

Substituting (1.39) and (1.40) in eq. (1.37), we have then proved for the estimated ground state energy that

$$\langle E \rangle \geq E_1 \qquad (1.41)$$

so that the variational method can always estimate an upper limit for the ground state energy in an arbitrary potential $V(x)$. Clearly, the more accurately the variational trial function, $f(x)$, is chosen, the closer the estimated variational energy $\langle E \rangle$ will be to the true ground state energy, E_1.

1.9 The variational method in an infinite square well

We illustrate the application of the variational method by considering the infinite square well of fig. 1.2. We know that the exact ground state in this case is given by $\psi_1(x) = (2/L)^{1/2}\sin(\pi x/L)$, but want to choose a different trial function, $f(x)$, as a simple test of the variational method. The trial function $f(x)$ must be chosen in this case so that its amplitude goes to zero at the two boundaries, $f(0) = f(L) = 0$, with the maximum expected in the centre, at $x = L/2$, and the function symmetric about $x = L/2$. The simplest polynomial function $f(x)$ to choose is the parabola

$$f(x) = x(L - x) \qquad (1.42)$$

for which we calculate that

$$Hf(x) = -\frac{\hbar^2}{2m}\frac{d^2}{dx^2}[x(L-x)] = \frac{\hbar^2}{m}$$

with the variational ground state energy then estimated as

$$\langle E \rangle = \frac{\int_0^L x(L-x)\frac{\hbar^2}{m}dx}{\int_0^L x^2(L-x)^2dx}$$

$$= \frac{\left(\frac{1}{6}\right)(\hbar^2/m)}{\left(\frac{1}{30}\right)L^2} = 0.12665\frac{\hbar^2}{mL^2} \tag{1.43}$$

This compares very well with the exact ground state energy calculated earlier using eq. (1.11) as $E_1 = 0.125\hbar^2/(mL^2)$, and demonstrates that the variational method works effectively, given a suitable choice of trial function. The accuracy with which we choose $f(x)$ can often be significantly improved by including a free parameter, say γ, in $f(x)$, and then calculating the variational energy $\langle E \rangle$ as a function of γ. When $d\langle E \rangle/d\gamma = 0$, we have generally minimised $\langle E \rangle$, and thereby achieved the best possible estimate of E_1 for the given $f(\gamma, x)$. This is described further in Appendix A, where we use the trial function $e^{-\gamma x}$ to estimate the electron ground state energy in the hydrogen atom.

As the ground state wavefunction is always even in a symmetric potential, choosing an odd function $g(x)$ allows an estimate of the first excited state energy. This is considered further in the problems at the end of this chapter, where we estimate the energy of the first excited state in the infinite square well potential, using a cubic function, $g(x)$, chosen as the simplest polynomial which is odd about the centre of the well, and zero at the edges.

1.10   The finite square well

As the finite square well proves useful for illuminating a wide range of problems in solid state physics, we complete this chapter by calculating the energy levels in a square quantum well of depth V_0 and width a, defined between $x = -a/2$ and $x = +a/2$, so that the potential is then symmetric about the origin (see fig. 1.8). We first review the conventional calculation of the confined state energies, which takes full advantage of the symmetry. We then present an alternative, less frequently seen derivation, which will prove useful later when we use a double quantum well to model bonding in molecules, and when we use an infinite array of quantum wells to model a periodic solid, using what is referred to as the Kronig–Penney (K–P) model.

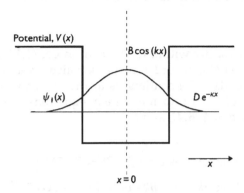

Figure 1.8 The thick solid line indicates a square well potential centred at the origin.
The thinner curve shows the position dependence, $\psi_1(x)$, of the ground
state wavefunction, while the thin horizontal line indicates the energy E_1
of the state.

We choose the zero of energy at the bottom of the quantum well so that,
within the well, Schrödinger's equation is given by

$$-\frac{\hbar^2}{2m}\frac{d^2\psi(x)}{dx^2} = E\psi(x) \qquad (1.44)$$

which has the general solution

$$\psi(x) = A\sin kx + B\cos kx \qquad |x| \leq a/2 \qquad (1.45)$$

where we have defined $k^2 = 2mE/\hbar^2$. Although classically a particle with
energy $E \leq V_0$ cannot penetrate into the barrier, there is a finite probability
of finding the particle there in quantum mechanics; Schrödinger's equation
in the barrier takes the form

$$-\frac{\hbar^2}{2m}\frac{d^2\psi(x)}{dx^2} + V_0\psi(x) = E\psi(x) \qquad (1.46)$$

which has the general solution

$$\psi(x) = Ce^{\kappa x} + De^{-\kappa x} \qquad x \geq a/2$$

$$= Fe^{\kappa x} + Ge^{-\kappa x} \qquad x \leq -a/2 \qquad (1.47)$$

with $\kappa^2 = 2m(V_0 - E)/\hbar^2$.

The allowed solutions of Schrödinger's equation are those which satisfy
the necessary boundary conditions, namely,

1 that the amplitude of $\psi \to 0$ as $x \to \pm\infty$, requiring $C = 0$ and $G = 0$
   (otherwise there would be an exponentially increasing probability of
   finding the particle at large $|x|$);

2   the wavefunction ψ and its derivative $d\psi/dx$ must also be continuous at all x.

This holds automatically within the well from eq. (1.45) and within the barrier from eq. (1.47); in order to be satisfied everywhere, we then require ψ and $d\psi/dx$ to be continuous at the well/barrier interfaces, at $x = \pm a/2$. This gives rise to four linear equations involving the four unknown parameters $A, B, D,$ and F:

$$\psi(a/2): \quad A\sin(ka/2) + B\cos(ka/2) = D\,e^{-\kappa a/2} \tag{1.48a}$$

$$\psi'(a/2): \quad Ak\cos(ka/2) - Bk\sin(ka/2) = -\kappa D\,e^{-\kappa a/2} \tag{1.48b}$$

$$\psi(-a/2): \quad -A\sin(ka/2) + B\cos(ka/2) = F\,e^{-\kappa a/2} \tag{1.48c}$$

$$\psi'(-a/2): \quad Ak\cos(ka/2) + Bk\sin(ka/2) = \kappa F\,e^{-\kappa a/2} \tag{1.48d}$$

These four equations can be solved directly, as we do below, to find the allowed energy levels, E_n, for states confined within the quantum well. However, because the potential is symmetric, it is easier to calculate separately the even and odd allowed states.

For the even states (fig. 1.8), $D = F$, and $\psi(x) = B\cos(kx)$ within the well, giving as boundary conditions at $x = a/2$

$$B\cos(ka/2) = D\,e^{-\kappa a/2} \tag{1.49a}$$

$$-Bk\sin(ka/2) = -D\kappa e^{-\kappa a/2} \tag{1.49b}$$

with two identical boundary conditions obtained at $x = -a/2$. Dividing (1.49b) by (1.49a), we obtain for the even states within the quantum well that

$$k\tan(ka/2) = \kappa$$

or

$$k\sin(ka/2) - \kappa\cos(ka/2) = 0 \tag{1.50}$$

We obtain the odd states by letting $D = -F$ and $\psi(x) = A\sin(kx)$ within the well, so that

$$k\cot(ka/2) = -\kappa$$

or

$$k\cos(ka/2) + \kappa\sin(ka/2) = 0 \tag{1.51}$$

The allowed energy levels are then determined by finding those values of $E = \hbar^2 k^2/2m$ for which either of the transcendental equations (1.50) or (1.51) can be satisfied.

We could have ignored the symmetry properties of the quantum well, allowing all confined states to have the general form given by eq. (1.45) in the well, and then found the allowed energy levels by directly solving the four linear equations in (1.48); that is, requiring that

$$
\begin{pmatrix}
\sin(ka/2) & \cos(ka/2) & -1 & 0 \\
k\cos(ka/2) & -k\sin(ka/2) & \kappa & 0 \\
-\sin(ka/2) & \cos(ka/2) & 0 & -1 \\
k\cos(ka/2) & k\sin(ka/2) & 0 & -\kappa
\end{pmatrix}
\begin{pmatrix}
A \\
B \\
D e^{-\kappa a/2} \\
F e^{-\kappa a/2}
\end{pmatrix}
=
\begin{pmatrix}
0 \\
0 \\
0 \\
0
\end{pmatrix}
\qquad (1.52)
$$

Non-trivial solutions of eq. (1.52) are obtained when the determinant of the 4×4 matrix is zero; it can be explicitly shown that the determinant is zero when

$$[k\sin(ka/2) - \kappa\cos(ka/2)][k\cos(ka/2) + \kappa\sin(ka/2)] = 0 \qquad (1.53)$$

which, not surprisingly, is just a combination of the separate conditions in eqs (1.50) and (1.51) for allowed even and odd states. When we multiply out the two terms in eq. (1.53) and use the standard trigonometric identities $\cos\theta = \cos^2(\theta/2) - \sin^2(\theta/2)$ and $\sin\theta = 2\cos(\theta/2)\sin(\theta/2)$, we obtain an alternative transcendental equation which must be satisfied by confined states in a square well, namely

$$(\kappa^2 - k^2)\sin ka + 2k\kappa\cos ka = 0$$

or

$$\cos ka + \tfrac{1}{2}(\kappa/k - k/\kappa)\sin ka = 0 \qquad (1.54)$$

This less familiar form of the conditions for allowed states in a finite square well potential will be very useful when investigating the allowed energy levels in a 'diatomic', or double quantum well in Chapter 2, and also when using the Kronig-Penney model for periodic solids in Chapter 3.

References

There are many introductory (and advanced) texts on quantum mechanics. A more detailed discussion of the topics considered here can be found for instance in:

Beiser, A. (2002) *Concepts of Modern Physics*, McGraw-Hill Inc., New York.
Eisberg, R. and R. Resnick (1985) *Quantum Physics of Atoms, Molecules, Solids, Nuclei, and Particles*, Second Edition, Wiley, New York.
Davies, P. C. W. and D. S. Betts (1994) *Quantum Mechanics*, Second Edition, Nelson Thornes, Cheltenham.

McMurry, S. M. (1993) *Quantum Mechanics*, Second Edition, Addison-Wesley, London.

Matthews, P. T. (1996) *Introduction to Quantum Mechanics*, McGraw-Hill Education, New York.

Schiff, L. I. (1968) *Quantum Mechanics*, Third Edition, McGraw-Hill, Tokyo.

Problems

1.1   Show that in a quantum well of depth V_0 and width a the energies of states of odd parity are given by $-k \cot(ka/2) = \kappa$, where $k^2 = 2mE/\hbar^2$ and $\kappa^2 = 2m(V_0 - E)/\hbar^2$.

1.2   Normalise the wavefunctions, $\psi_n(x) = a_n \sin(n\pi x/L)$, of the infinite square well, for which $V(x) = 0$, for $0 < x < L$, and $= \infty$ otherwise. Show that the wavefunctions are orthogonal to each other, that is,

$$\int_0^L \psi_n^*(x)\psi_m(x)\,\mathrm{d}x = \delta_{mn}$$

1.3   A trial function, $f(x)$, differs from the ground state wavefunction, $\psi_1(x)$, by a small amount, which we write as

$$f(x) = \psi_1(x) + \varepsilon u(x)$$

where $\psi_1(x)$ and $u(x)$ are normalised, and $\varepsilon \ll 1$. Show that $\langle E \rangle$, the variational estimate of the ground state energy E_1, differs from E_1 only by a term of order ε^2, and find this term. [This shows that the relative errors in the calculated variational energy can be considerably smaller than the error in the trial function used.]

1.4   Consider an infinite square well between $-L/2$ and $+L/2$.

a   Use the variational method to estimate the ground state energy in this well assuming $f(x) = (L/2)^n - x^n$, where n is an even integer, ≥ 2. Comment why the function becomes an increasingly unsuitable starting function with increasing n.

b   Justify the choice of the cubic function $g(x) = (2x/L) - (2x/L)^3$ to estimate the energy of the first excited state. Use $g(x)$ to estimate E_2 and compare your result with the exact solution.

c   Suggest a suitable polynomial form for the variational function which might be chosen to estimate the energy of the second and higher excited states.

1.5   Consider a particle moving in the one-dimensional harmonic oscillator potential, $V(x) = \frac{1}{2}kx^2$. By using the trial function, $f(x) = \exp(-\alpha x^2)$, estimate the ground state energy of the harmonic oscillator. We can use $g(x) = x\exp(-\beta x^2)$ as a trial function to estimate the lowest state of odd parity, that is, the first excited state. Estimate this energy.

Chapter 2

Bonding in molecules and solids

2.1 Introduction

Many trends in the properties of solids follow directly from trends in the properties of the constituent atoms. The semiconductors germanium (Ge), gallium arsenide (GaAs) and zinc selenide (ZnSe) are all formed from atoms in the same row of the periodic table: they all have the same crystal structure and approximately the same lattice constant, but the fundamental band gap increases on going from the covalently bonded group IV semiconductor Ge to the polar III–V compound GaAs, and again on going to the even more polar II–VI compound ZnSe. Silicon (Si) is fourfold-coordinated, with four nearest neighbour atoms in almost all of the compounds which it forms, while nitrogen (N) is generally three-coordinated, as in ammonia (NH_3) or silicon nitride (Si_3N_4), where each Si has four N and each N three nearest Si neighbours.

The observation of such properties and their classification through the Periodic Table of the elements predated the Schrödinger equation by over fifty years, but it took the development of quantum mechanics to first explain the structure of the periodic table, and the trends in atomic properties with increasing atomic number. It took longer still to explain how the atomic trends give rise to the observed trends in the chemical and physical properties of matter.

Some of the observed properties, such as high temperature superconductivity, have still to be fully understood, but there have been many significant advances in recent years in the development of both approximate and first principles methods to explain and predict a wide range of material properties, each of which is the subject in its own right of major text books and review papers.

We are largely concerned in this chapter with understanding the origins of chemical bonding in molecules and solids: how, as we bring atoms closer together the atomic energy levels play a significant and predictable role in determining the electronic energy levels of the resultant molecule or solid. We illustrate this by first taking the square well as a prototype atom and investigating analytically the evolution of the energy level spectrum as the separation, b, between two square wells is decreased to give a diatomic, or double square well potential. We then apply a variational method to the same problem, showing that linear combinations of the 'atomic orbitals' (i.e. wavefunctions) of the isolated square wells enable a good description of the double quantum well energy level spectrum up to surprisingly small values of the well separation, b. The square well illustrates many, but not all, properties of atomic bonding, as it obviously omits factors such as nucleus–nucleus repulsion. Having used the square well to establish the applicability of a variational method, we then consider the hydrogen molecule, H_2, followed by an ionic molecule, chosen to be LiH, to illustrate the effects of bonding between dissimilar atoms. This analysis provides a model which is widely applicable, explaining bonding trends for instance in crystalline semiconductors and insulators, disordered solids, large polymer chains and small molecules. It should however be remarked that the bonding model developed here, based on linear combinations of atomic orbitals, at best, only partly explains the origins of and trends in bonding in metals: a fuller understanding requires consideration of the evolution from isolated energy levels in atoms to energy bands in solids, which will be discussed further in Chapter 3. A particularly good and more extended description of much of the material in this chapter can be found in Harrison (1980, 2000).

2.2 Double square well potential

The solid line in fig. 2.1 shows the potential $V(x)$ associated with two square quantum wells, each of width a and depth V_0, separated by a barrier

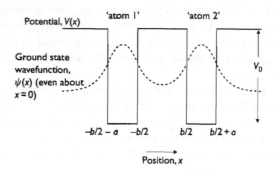

Figure 2.1 The solid line shows two square quantum wells of width a and depth V_0 separated by a barrier of width b. This potential is chosen to model the interaction between two 'atoms' whose 'nuclei' are a distance $a + b$ apart, and with each 'atom' modelled by a square well potential. The dashed line illustrates the wavefunction for the lowest eigenstate, which is symmetric about $x = 0$.

of width b. We choose the origin at the centre of this barrier, so that the potential is then symmetric about $x = 0$, with the right-hand well then between $x = b/2$ and $b/2+a$, and the left-hand well between $x = -(b/2+a)$ and $-b/2$. With this symmetry, the double well wavefunctions will be either even or odd about $x = 0$. The wavefunction for the lowest symmetric state is illustrated by the dashed line in fig. 2.1, and can be written down in terms of four unknown parameters A, B, C and D:

1   Within the central barrier, Schrödinger's equation takes the form of eq. (1.46):

$$-\frac{\hbar^2}{2m}\frac{d^2\psi}{dx^2} + V_0\psi = E\psi \tag{2.1}$$

for which the even solution is

$$\psi(x) = A(e^{\kappa x} + e^{-\kappa x}) \qquad -\frac{b}{2} < x < \frac{b}{2} \tag{2.2}$$

with $\kappa^2 = 2m(V_0 - E)/\hbar^2$.

2   Within the right-hand well, Schrödinger's equation is given by eq. (1.44)

$$-\frac{\hbar^2}{2m}\frac{d^2\psi}{dx^2} = E\psi \tag{2.3}$$

and we choose as our general solution

$$\psi(x) = B\cos\left(k\left(x - \frac{a+b}{2}\right)\right) + C\sin\left(k\left(x - \frac{a+b}{2}\right)\right)$$
$$\frac{b}{2} < x < \left(a + \frac{b}{2}\right) \tag{2.4}$$

where $k^2 = 2mE/\hbar^2$, with the phase of the sine and cosine functions chosen so that they are, respectively, odd and even about the well centre, $(a + b)/2$.

3   Schrödinger's equation has the same form in the right hand as in the central barrier; and in order that $\psi \rightarrow 0$ as $x \rightarrow \infty$, we choose

$$\psi(x) = De^{-\kappa(x-(b/2+a))} \qquad x > \left(a + \frac{b}{2}\right) \tag{2.5}$$

Because of the symmetry, the wavefunctions in the left-hand well and barrier depend on the same unknown coefficients B, C, and D.

We require for allowed solutions of Schrödinger's equation that the wavefunction ψ and its derivative $d\psi/dx$ be continuous at the well/barrier interfaces; namely,

$$\psi(b/2): \quad A(e^{\kappa b/2} + e^{-\kappa b/2}) = B\cos(ka/2) - C\sin(ka/2) \tag{2.6a}$$

$$\psi'(b/2): \quad \kappa A(e^{\kappa b/2} - e^{-\kappa b/2}) = kB\sin(ka/2) + kC\cos(ka/2) \tag{2.6b}$$

$$\psi(a+b/2): \quad D = B\cos(ka/2) + C\sin(ka/2) \tag{2.6c}$$

$$\psi'(a+b/2): \quad -\kappa D = -kB\sin(ka/2) + kC\cos(ka/2) \tag{2.6d}$$

We could find the conditions for allowed energy levels by solving the 4×4 determinant involving the four unknowns A, B, C, and D, as was done with eq. (1.52) in Chapter 1. Alternatively (or equivalently) we can first use (2.6c) and (2.6d) to determine B and C in terms of D:

$$C = D\left(\sin(ka/2) - \left(\frac{\kappa}{k}\right)\cos(ka/2)\right) \tag{2.7a}$$

$$B = D\left(\cos(ka/2) + \left(\frac{\kappa}{k}\right)\sin(ka/2)\right) \tag{2.7b}$$

and then by substituting the values for B and C in (2.6a) and (2.6b), and using the double angle identities, we obtain

$$A(e^{\kappa b/2} + e^{-\kappa b/2}) = D((\kappa/k)\sin(ka) + \cos(ka)) \tag{2.8a}$$

$$A(e^{\kappa b/2} - e^{-\kappa b/2}) = D((k/\kappa)\sin(ka) - \cos(ka)) \tag{2.8b}$$

Dividing (2.8b) by (2.8a) we find

$$\tanh\left(\frac{\kappa b}{2}\right) = \frac{(k/\kappa)\sin(ka) - \cos(ka)}{(\kappa/k)\sin(ka) + \cos(ka)} \tag{2.9a}$$

which can be rewritten as

$$\left(\kappa^2 \tanh\left(\frac{\kappa b}{2}\right) - k^2\right)\sin(ka) + k\kappa\left(\tanh\left(\frac{\kappa b}{2}\right) + 1\right)\cos(ka) = 0 \tag{2.9b}$$

A similar analysis shows that the energy levels for the double well states of odd parity are found as solutions of

$$\left(\kappa^2 \coth\left(\frac{\kappa b}{2}\right) - k^2\right)\sin(ka) + k\kappa\left(\coth\left(\frac{\kappa b}{2}\right) + 1\right)\cos(ka) = 0 \tag{2.10}$$

We consider means of solving eqs (2.9) and (2.10) in the problems at the end of this chapter. Even without solving exactly, we can deduce several important results concerning coupled quantum wells and molecular bonding from these equations:

(1) *Isolated well limit*: As $b \to \infty$, and the square quantum wells become widely separated from each other, we expect the energy levels to approach those for an isolated well. This is confirmed by noting that both $\tanh(\kappa b)$ and $\coth(\kappa b) \to 1$, as $b \to \infty$, so that both eq. (2.9) and eq. (2.10) then take the form

$$(\kappa^2 - k^2)\sin(ka) + 2\kappa k \cos(ka) = 0 \qquad (2.11)$$

which is just the condition (eq. (1.54)) to determine the energy levels of an isolated quantum well.

(2) *Bonding and anti-bonding levels*: Conversely, we see from eqs (2.9) and (2.10) that as the interwell separation b decreases the coupled well energy levels evolve continuously from the isolated single quantum well levels. This is illustrated in fig. 2.2(a) and (b) where we plot the evolution of energy levels in a shallow well and a deep well as a function of b. As b decreases, the doubly degenerate levels start to move apart, one going up and the other down in energy, with the splitting increasing for decreasing separation b. This splitting explains the origins of chemical bonding. Two electrons (of opposite spin) can occupy each energy level in an atom

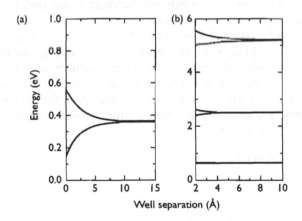

Figure 2.2 Variation of allowed energy levels as a function of the separating barrier width b for (a) two coupled 'shallow' quantum wells (each of width $a = 6$ Å and depth $V_0 = 1.0$ eV), and (b) two 'deep' quantum wells (also of width $a = 6$ Å, but of depth $V_0 = 6.0$ eV). As the separation $b \to \infty$, each of the shallow wells has one confined energy level, while each of the deeper wells in (b) has three confined states.

or molecule, with the lowest energy levels being filled first. If each of the quantum wells or 'atoms' in fig. 2.2(a) has one electron in the level bound in the quantum well, then the overall energy will be reduced by forming the coupled quantum well, where the two electrons can occupy the lowest energy level, as in the diatomic hydrogen molecule, H_2. This level is referred to as the bonding level. By contrast, if there are already two electrons in the highest filled level of each isolated well or atom, as for helium (He), the second element in the Periodic Table, it will cost energy to form a molecule, with two of the electrons going into the lower (bonding) level and the other two into the upper (anti-bonding) level. Hence He gas is made up of isolated He atoms rather than He_2 or more complicated molecules.

(3) *Core and valence levels*: Returning to the deep well in fig. 2.2(b), we see for moderate b that the splitting between the highest energy levels is considerably larger than is the case for the lower energy levels. This can be understood from eqs (2.9) and (2.10) if we note that the magnitude of $\kappa = \{2m(V_0 - E)/\hbar^2\}^{1/2}$ increases as the energy decreases, going towards the bottom of the quantum well. In this situation, $\tanh(\kappa b)$ and $\coth(\kappa b)$ are much closer to 1 for the lower energy levels than for those nearer to the well maximum, and so the deep states are far less perturbed from their isolated well values compared to the higher levels. The same is true in molecules and solids where the deeper energy levels, referred to as core levels, are largely unperturbed and do not take part in bonding. Hence, despite having different numbers of core electrons, gaseous flourine (F), chlorine (Cl) and iodine (I) all exist as diatomic molecules, F_2, Cl_2 and I_2, as all have the same number of valence electrons.

(4) *Linear combinations of atomic orbitals*: Figure 2.3(a) and (b) show the wavefunctions (solid lines) for the shallow-well bonding and anti-bonding energy levels at selected values of interwell separation, b. It can be seen even for small b that the wavefunctions are virtually indistinguishable from symmetric and anti-symmetric combinations of the isolated quantum well functions (indicated by the dotted lines). We may write the coupled well wavefunction $\psi(x)$ as

$$\psi_s(x) = \alpha(\phi_L(x) + \phi_R(x))$$
$$\psi_a(x) = \beta(\phi_L(x) - \phi_R(x))$$
(2.12)

for the symmetric and anti-symmetric cases, respectively, where $\phi_L(x)$ and $\phi_R(x)$ are eigenstates of an isolated left-hand and right-hand well respectively, and $\alpha = \beta = 1/\sqrt{2}$ for large separation. We see from fig. 2.3 that this linear combination of isolated well wavefunctions (i.e. 'atomic orbitals') should act as a good variational guess at the molecular wavefunctions up to relatively small well separations b. This is indeed confirmed by fig. 2.4(a) and (b), which compare the exact double well energy levels of fig. 2.2 (solid

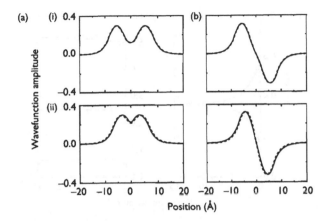

Figure 2.3 The wavefunctions of the (a) symmetric and (b) anti-symmetric confined
state in the coupled shallow wells of fig. 2.2(a), calculated for central barrier
widths of (i) $b = 5\,\text{Å}$ and (ii) $b = 2\,\text{Å}$. The solid lines show the exact wavefunctions, while the dashed lines are a variational estimate assuming each
wavefunction to be a linear combination of isolated well wavefunctions.

lines) with the energy levels calculated using eq. (1.37) and the variational
wavefunctions of eq. (2.12) (dotted lines): agreement between the two
remains good to small well separation b. This ability to use isolated atomic
wavefunctions as basis states in variational calculations explains in large
part why atomic properties play such a major role in determining trends
in the observed chemical and physical properties of molecules and solids.

We note that the variational method does break down here as $b \to 0$,
particularly for the excited states in the deeper well (fig. 2.4(b)). This is not
surprising as at $b = 0$, eqs (2.9) and (2.10) reduce to

$$k \tan(ka) = \kappa$$

and

$$k \cot(ka) = -\kappa \tag{2.13}$$

respectively, which are just the equations determining the even and odd
energy levels in a well of width $2a$.

In summary, through the example of the double quantum well, we have
shown

1   how molecular energy levels evolve continuously from those of
    isolated atoms, as the atoms are brought closer together;
2   how repulsion between energy levels on neighbouring atoms can lead
    to the formation of bonding and anti-bonding states;

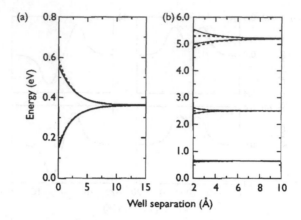

Figure 2.4 Comparison of the exact double well energy levels of fig. 2.2 (solid lines) with the energy levels calculated using the variational method, where the double well variational functions are taken to be a linear combination of exact isolated well wavefunctions.

3   that core levels play no part in molecular bonding, and
4   that the molecular wavefunctions can often be well approximated by linear combinations of atomic orbitals.

We turn in the next section to look more closely at the hydrogen molecule, H_2, and then at molecules such as LiH, formed from two atoms which are not identical.

2.3 The hydrogen molecule, H_2

For an isolated hydrogen atom, the Schrödinger equation is given by

$$H\psi(r) = \left(-\frac{\hbar^2}{2m}\nabla^2 - \frac{e^2}{4\pi\varepsilon_0 r}\right)\psi(r) = E\psi(r) \qquad (2.14)$$

with the ground state energy, E_H, shown in Appendix A to be equal to $-13.6\,\text{eV}$ and the electron wavefunction given by $\phi(r) = C\exp(-r/a_0)$, referred to as the ground state atomic orbital. If we bring two hydrogen atoms, a and b, together to form a molecule, then the isolated atomic orbitals, $\phi_a(r)$ and $\phi_b(r)$, are no longer eigenfunctions of the combined Hamiltonian

$$H\psi(r) = \left(-\frac{\hbar^2}{2m}\nabla^2 - \frac{e^2}{4\pi\varepsilon_0 r_a} - \frac{e^2}{4\pi\varepsilon_0 r_b} + \frac{e^2}{4\pi\varepsilon_0 r_{12}}\right)\psi(r) \qquad (2.15)$$

where we take r_a to be the electron distance from nucleus a, r_b from nucleus b, and r_{12} the electron–electron separation. Our solution of eq. (2.15) uses what is referred to as the one-electron approximation, where we assume that each of the electrons effectively sees an average potential due to the other electron. We also assume, as we did above, that the molecular wavefunction can be written as a linear combination of the atomic orbitals, which we now write as

$$\psi(r) = \alpha \phi_a(r) + \beta \phi_b(r) \tag{2.16}$$

Schrödinger's equation then takes the form

$$H(\alpha \phi_a(r) + \beta \phi_b(r)) = E(\alpha \phi_a(r) + \beta \phi_b(r)) \tag{2.17}$$

and the parameters α and β need to be determined in order to find the allowed molecular energy levels, E_i. We can determine the ground state energy by applying the variational method as described in Chapter 1 but, in order to estimate the magnitude of some of the integrals involved, we break the calculation down into several steps. We first multiply eq. (2.17) from the left by $\phi_a^*(r)$ and integrate over all space:

$$\alpha \underbrace{\int d^3r(\phi_a^*(r)H\phi_a(r))}_{\text{I}} + \beta \underbrace{\int d^3r(\phi_a^*(r)H\phi_b(r))}_{\text{II}}$$

$$= \alpha E \underbrace{\int d^3r(\phi_a^*(r)\phi_a(r))}_{\text{III}} + \beta E \underbrace{\int d^3r(\phi_a^*(r)\phi_b(r))}_{\text{IV}} \tag{2.18}$$

We must next estimate the magnitude of the four terms I, II, III and IV in eq. (2.18). We note that the first term can be split into two parts:

$$\text{I}: \int d^3r \phi_a^*(r) \left(-\frac{\hbar^2}{2m}\nabla^2 - \frac{e^2}{4\pi\varepsilon_0 r_a} \right) \phi_a(r)$$

$$+ \int d^3r \phi_a^*(r) \left(-\frac{e^2}{4\pi\varepsilon_0 r_b} + \frac{e^2}{4\pi\varepsilon_0 r_{12}} \right) \phi_a(r) \tag{2.19}$$

where the first part is just the Schrödinger equation for an electron in an isolated hydrogen atom, and is of magnitude E_H, while the second part involves an integral over the sum of the attractive electron–nucleus b and repulsive electron–electron potentials, which should approximately cancel, leaving the value of the second term close to zero.

Because we have normalised the wavefunctions, the integral in the term III is equal to one, that is $\int d^3r \phi_a^*(r)\phi_a(r) = 1$, while the term IV describes

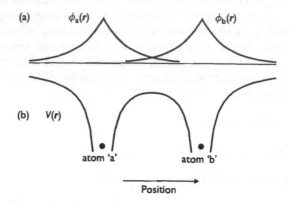

(a) $\phi_a(r)$ $\phi_b(r)$

(b) $V(r)$

atom 'a'   atom 'b'

Position

Figure 2.5 Variation of (a) the two isolated atomic wavefunctions, ϕ_a and ϕ_b, and (b) the electron potential, $V(r)$ along the axis joining the two hydrogen nuclei in a H_2 molecule.

the overlap between atomic orbitals centred on atoms a and b (see fig. 2.5). We introduce the overlap parameter S,

$$\int d^3 r \phi_a^*(r)\phi_b(r) = S \tag{2.20}$$

where $0 < S < 1$, and we expect for moderate overlap that S is generally significantly less than 1.

Finally, we consider the term II which describes the interaction via the Hamiltonian H between an atomic orbital on site a and one on site b. We introduce the parameter U such that

$$\int d^3 r \phi_a^*(r)H\phi_b(r) = U \tag{2.21}$$

where, from fig. 2.5, $U < 0$, and is a measure of the strength of the interaction between an electron on a and an electron on b, with the magnitude of U increasing with decreasing separation between the two atoms. Equation (2.18) can then be rewritten as

$$\alpha E_H + \beta U = \alpha E + \beta E S \tag{2.22}$$

If we multiply eq. (2.17) from the left by $\phi_b^*(r)$ and then integrate over all space, we get a second equation involving α and β, namely

$$\alpha U + \beta E_H = \alpha E S + \beta E \tag{2.23}$$

We can rewrite eqs (2.22) and (2.23) as

$$\alpha(E_H - E) + \beta(U - ES) = 0$$
$$\alpha(U - ES) + \beta(E_H - E) = 0 \tag{2.24}$$

These simultaneous equations can be solved for an arbitrary value of S, but the solution is simplifed if we assume the overlap $S \approx 0$, in which case the allowed energy levels are at

$$E = E_H \pm U \tag{2.25}$$

with the molecular variational wavefunctions then given by

$$\psi_\pm(r) = \frac{1}{\sqrt{2}}(\phi_a(r) \pm \phi_b(r)) \tag{2.26}$$

We note that as the amplitude of the atomic orbitals ϕ_a and ϕ_b decays exponentially with increasing distance, the overlap interaction U and hence the splitting between the bonding and anti-bonding levels also decreases exponentially with increasing separation between atoms a and b, as was also observed for the double quantum well energy levels in fig. 2.2(a) and (b).

The sum of the binding energies of two isolated hydrogen atoms is $2|E_H|$. In a H_2 molecule, the two electrons go into the lowest energy level, where, from the simple model here, the binding energy becomes $2(|E_H| + |U|)$, as illustrated in fig. 2.6, so that the total binding energy is then increased by $2|U|$ by forming H_2. This explains why hydrogen normally exists as the diatomic molecule H_2 rather than as isolated atoms.

It might appear from the above analysis and eq. (2.25) that the lowest energy state would be achieved when the magnitude of U is maximised, and the two atomic nuclei are at the same point. However, the estimate of the total binding energy here double-counts the electron–electron repulsion, while ignoring the nuclear–nuclear repulsion. At moderate separations, these two errors approximately cancel, but as the separation decreases the distance between the two nuclei becomes smaller than the average electron–electron distance, leading to an underestimate of the repulsion energy, and hence to a maximum binding energy at finite separation, as illustrated in fig. 2.7.

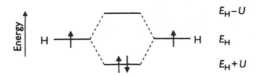

Figure 2.6 Schematic energy level diagram, illustrating how the interaction between two hydrogen atoms, each with isolated orbital energy, E_H, gives rise to a doubly filled bonding state at $E_H + U$ and empty anti-bonding state at $E_H - U$.

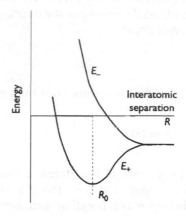

Figure 2.7 Schematic variation of the bonding (labelled E_+) and lowest anti-bonding energy level (labelled E_-) in a hydrogen molecule, H_2, as a function of interatomic separation, R. The equilibrium separation is at $R = R_0$.

2.4   The diatomic LiH molecule

We turn now to consider bonding in a diatomic molecule made up of two different atoms, such as LiH. We can go through a similar variational argument as before, guessing that the lowest energy molecular wavefunctions are a linear combination of a hydrogen atomic orbital, $\phi_H(r)$, and a lithium atomic orbital, $\phi_{Li}(r)$:

$$\psi_i(r) = \alpha\phi_H(r) + \beta\phi_{Li}(r) \tag{2.27}$$

If we now multiply eq. (2.27) from the left by $\phi_H^*(r)$ or $\phi_{Li}^*(r)$ and integrate over all space, we again get two linear equations which must be satisfied

$$\alpha(E_H - E) + \beta U = 0$$
$$\alpha U + \beta(E_{Li} - E) = 0 \tag{2.28}$$

The bonding E_+ and anti-bonding E_- energy levels are then calculated to lie at

$$E_+ = \frac{E_H + E_{Li}}{2} - \sqrt{\frac{(E_H - E_{Li})^2}{4} + U^2}$$

and

$$E_- = \frac{E_H + E_{Li}}{2} + \sqrt{\frac{(E_H - E_{Li})^2}{4} + U^2} \tag{2.29}$$

respectively, as illustrated in fig. 2.8.

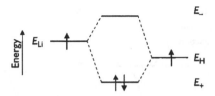

The energy gap, E_g, between the highest occupied molecular orbital (known as the HOMO) and the lowest unoccupied molecular orbital (LUMO) is then given by

$$E_g = 2\sqrt{U^2 + \frac{(E_H - E_{Li})^2}{4}}$$

or

$$E_g^2 = (2U)^2 + (E_H - E_{Li})^2 \tag{2.30}$$

We rewrite this as

$$E_g^2 = E_h^2 + C^2 \tag{2.30a}$$

to emphasise that there are two contributions to the energy gap, the covalent bonding contribution $E_h^2 = (2U)^2$ (where the subscript 'h' denotes homopolar), and the ionic contribution, $C^2 = (E_H - E_{Li})^2$, due to the difference in atomic orbital energy, which is often described by chemists in terms of the difference in electronegativity of the two atoms. For any bond, we can then define the bond covalency, α_c, and polarity, α_p, by

$$\alpha_c = \frac{2U}{\sqrt{(2U)^2 + (E_a - E_c)^2}} = \frac{E_h}{E_g}$$

and

$$\alpha_p = \frac{|E_a - E_c|}{\sqrt{(2U)^2 + (E_a - E_c)^2}} = \frac{C}{E_g} \tag{2.31}$$

where E_a and E_c refer to the orbital energies of the two atoms. We then have for all bonds that

$$\alpha_c^2 + \alpha_p^2 = 1 \tag{2.32}$$

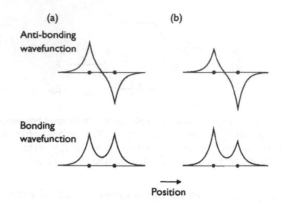

Figure 2.9 (a) In the purely covalent hydrogen molecule, the bonding and anti-bonding wavefunctions are both shared equally between the two hydrogen atoms. (b) In a polar molecule, the amplitude of the bonding state shifts towards the more electronegative site, with the anti-bonding state shifting towards the less electronegative site (where electrons are less tightly bound).

so that $\alpha_c = 1$ for a purely covalent bond, such as in H_2 or a Si crystal. More interestingly, there is a continuous transition from covalent to ionic bonding. As the ionicity increases, the bonding state shifts predominantly towards the more electronegative site, that is, the site with more tightly bound valence electrons, which is referred to as the anion. At the same time, the anti-bonding state shifts to the less electronegative site, referred to as the cation. This is illustrated in fig. 2.9. The values of α and β can be calculated in eq. (2.27) to find that

$$\psi_+ = \sqrt{\frac{1+\alpha_p}{2}}\phi_a + \sqrt{\frac{1-\alpha_p}{2}}\phi_c$$

while

$$\psi_- = \sqrt{\frac{1-\alpha_p}{2}}\phi_a - \sqrt{\frac{1+\alpha_p}{2}}\phi_c \tag{2.33}$$

where ϕ_a and ϕ_c refer to orbitals on the anion and cation with energies E_a and E_c, respectively. There is, therefore, a continuous transition from purely covalent bonding, as in Si or Ge, to polar bonding, as in GaAs, and to ionic bonding, as in sodium chloride, NaCl – common salt – where most of the binding energy is associated with the electrostatic attraction due to the transfer of one electron from each sodium to each chlorine atom, giving Na^+Cl^-. We discuss in the next sections how trends in the electronic properties of tetrahedrally bonded semiconductors can be understood in terms of trends in the covalent and ionic contributions to their bonding.

2.5   Tetrahedral bonding in Si crystals

The elements in the right hand columns of the Periodic Table have four valence atomic orbitals: one s-state which is spherically symmetric, and three p-states, which can be described as having x-, y- and z-like symmetry, as described in Appendix B. Such orbitals are illustrated in fig. 2.10, where we have drawn surfaces on which the magnitude of the amplitude of each atomic orbital is constant.

The crystal lattice structure of Si is shown in fig. 2.11. Each Si atom has four nearest neighbours, with the neighbours forming a tetrahedron about the central atom. The Si lattice in fig. 2.11 is often described as being made up of two interpenetrating face-centred-cubic (FCC) lattices: one of these can be reasonably clearly seen by marking the atoms on each corner of the cube shown and also on the face centres. This description is of little relevance here but will be useful later when considering polar semiconductors, such as GaAs, and when considering the crystal band structure in more detail in Chapter 3.

To get a simple understanding of bonding in a tetrahedral semiconductor such as Si, we convert from the atomic orbital basis of one s and three p states to a basis of directional hybrid orbitals, referred to as sp^3 orbitals. Hybrid states are formed by taking linear combinations of the atomic orbitals on an atom, and can be chosen to give highly directional orbitals whose amplitude is maximised along the direction to a specific nearest neighbour. Consider for instance an s-state, ϕ_s and a p$_x$-state, ϕ_x,

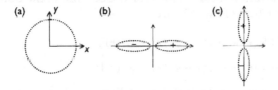

Figure 2.10 Surface of constant amplitude for (a) an s-state, (b) a p$_x$-state, and (c) a p$_y$-state. The third p state (not illustrated) is directed along the z-direction (out of the page).

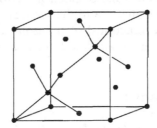

Figure 2.11 Crystal lattice structure of Si. Each Si atom has four nearest neighbours, with the neigbours forming a tetrahedron about the given atom. (From H. P. Myers (1997) *Introductory Solid State Physics*, 2nd edn.)

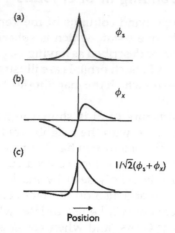

Figure 2.12 A hybrid orbital (c) formed as a linear combination of an s-state (a) and a p_x-state (b) can have greater amplitude along the x-direction than that of either of the states from which it is formed.

whose amplitudes vary along the x-axis as shown in fig. 2.12(a) and (b). Adding the two together, as $1/\sqrt{2}(\phi_s(r) + \phi_x(r))$, gives the hybrid state shown in fig. 2.12(c), which has a maximum amplitude for $x > 0$ that is larger than that of either of the constituent atomic orbitals.

The four relevant hybrid orbitals for the Si atom in the bottom corner of the crystal structure of fig. 2.11 are

$$\psi_1(r) = \tfrac{1}{2}[\phi_s(r) + \phi_x(r) + \phi_y(r) + \phi_z(r)]$$
$$\psi_2(r) = \tfrac{1}{2}[\phi_s(r) + \phi_x(r) - \phi_y(r) - \phi_z(r)]$$
$$\psi_3(r) = \tfrac{1}{2}[\phi_s(r) - \phi_x(r) + \phi_y(r) - \phi_z(r)]$$
$$\psi_4(r) = \tfrac{1}{2}[\phi_s(r) - \phi_x(r) - \phi_y(r) + \phi_z(r)]$$

(2.34)

An isolated Si atom has four valence electrons, two in the atomic s state, at energy E_s, and two in p states at a higher energy, E_p. The hybrid orbitals are not eigenstates (wavefunctions) of the isolated atom, but instead are at an energy E_h, which can be calculated, for example, for the hybrid state $\psi_1(r)$ as

$$E_h = \int d^3r\, \psi_1^*(r) H_0 \psi_1(r)$$
$$= \frac{1}{4} \int d^3r [\phi_s^*(r) + \phi_x^*(r) + \phi_y^*(r) + \phi_z^*(r)]$$
$$\times H_0[\phi_s(r) + \phi_x(r) + \phi_y(r) + \phi_z(r)]$$

(2.35)

where H_0 is the Hamiltonian for the isolated atom. Because the s- and p-orbitals are allowed energy levels of the isolated atom and are orthogonal to each other, most of the terms in eq. (2.35) disappear and we are left with

$$E_h = \frac{1}{4} \int d^3 r [\phi_s^*(r) H_0 \phi_s(r)$$
$$+ \phi_x^*(r) H_0 \phi_x(r) + \phi_y^*(r) H_0 \phi_y(r) + \phi_z^*(r) H_0 \phi_z(r)]$$
$$= \frac{1}{4}(E_s + 3E_p) \tag{2.36}$$

The energy of four electrons in the lowest available atomic states was $2E_s + 2E_p$, while that of four electrons, one in each of the four sp^3 hybrids is $E_s + 3E_p$, as illustrated in fig. 2.13. It therefore costs energy to form the directional sp^3 hybrids but, once they are formed, each hybrid interacts strongly with one hybrid on a neighbouring atom, to form filled bonding and empty anti-bonding states, thereby gaining a bonding energy of $4|U|$ per atom, where U is the hybrid interaction energy, and a net increase in the binding energy per atom of $4|U| + |E_p| - |E_s|$ in tetrahedrally bonded Si. Of course, there are many more interactions between the hybrid orbitals than the one we have just focussed on here betweeen two hybrids pointing towards each other. The interactions which we have ignored broaden the bonding levels into a filled valence band of states, and the anti-bonding levels into the empty conduction band. Nevertheless, the hybrid orbital picture presented here provides a convincing explanation for the crystal structure of tetrahedrally bonded semiconductors, and can also provide insight into trends in a variety of semiconductor properties, some of which we discuss in more detail below.

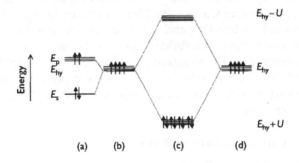

Figure 2.13 (a) An isolated Si atom has two electrons in an s-like valence level (at energy E_s here) and two electrons in p states (at energy E_p). (b) It costs energy to form four hybrid orbitals, each at energy E_{hy}, and each containing one electron. However each hybrid orbital can then interact strongly with a hybrid orbital on a neighbouring Si atom (d) to form a strongly bonding state at energy $E_{hy} - |U|$ (c).

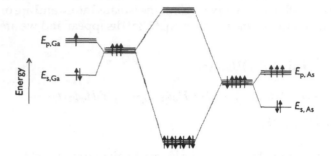

Figure 2.14 Tetrahedral bonding in GaAs can be explained through the formation of sp^3 hybrid orbitals on each Ga and As site (left- and right-hand side of figure respectively). Each Ga sp^3 hybrid then overlaps and interacts strongly with an sp^3 hybrid on a neighbouring As atom, to give strong polar bonding in GaAs.

2.6 Bonding in gallium arsenide

The crystal structure of GaAs is similar to that of Si, shown in fig. 2.11. The gallium (Ga) atoms occupy one of the FCC lattices which make up the crystal structure, with the arsenic (As) atoms on the other FCC lattice, so that now each Ga has four As nearest neighbours, and each As four Ga neighbours. An isolated As atom has five valence electrons, two in s-states, and three in p-states, while an isolated Ga has three valence electrons, two in the s level and one in a p level, as illustrated in fig. 2.14. We can again form hybrid orbitals on the Ga and As atoms, with the Ga hybrids lying above the As hybrids, because As, with tighter bound valence states, is more electronegative than Ga. Just as bonding in Si is akin to that in H$_2$, so we can compare GaAs with LiH, as shown in fig. 2.8, where the splitting between the bonding and anti-bonding levels has both a covalent contribution, due to the interhybrid interaction, U, and a polar contribution due to the difference in electronegativity. We note however that there is little net charge transfer between the Ga and As sites in polar-bonded GaAs, as the As atoms contribute in any case $\frac{5}{4}$ electrons and the Ga atoms $\frac{3}{4}$ of an electron to each bond.

2.7 Trends in semiconductors

The trends in band gap with covalency and ionicity predicted by eq. (2.30) are generally observed across the full range of semiconducting and insulating materials. Figure 2.15 shows as an example several rows and columns of the Periodic Table from which the sp^3-bonded, fourfold-coordinated semiconductors are formed. The Periodic Table is discussed in more detail in Appendix B. The number at the top of each column in fig. 2.15 indicates

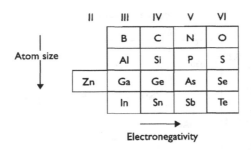

Figure 2.15 Selected elements from columns II to VI of the Periodic Table. The number at the top of each column indicates the number of valence electrons which the atoms in that column can contribute to bonding. Tetrahedral bonding can occur in group IV, III–V, and II–VI compounds, as the average number of valence electrons per atom is four in each of these cases.

the number of valence electrons which the given atom can contribute to bonding. The electronegativity tends to increase as we move along each row towards the right-hand end, due to increasingly large atomic orbital binding energies. The covalent radius (atomic size) is relatively constant within each row, but increases on going down to lower rows, because of the extra core electrons in the lower rows. The electronegativities also tend to be larger at the top of the Periodic Table than in lower rows because, with the increase in core radius in the lower rows, the electrons are less tightly bound to the nuclei.

As the magnitude of the covalent interaction, (U or E_h in eq. (2.30)) decreases with increasing atomic separation, we can predict that the band gap will decrease going down the series of purely covalent group IV semiconductors, from diamond (C) through Si and Ge to β-tin (Sn). We likewise expect it to decrease going down the series of polar III–V compounds, aluminium phosphide (AlP) through GaAs to indium antimonide (InSb). On the other hand, if we take a set of tetrahedral semiconductors from the same row of the Periodic Table (where the covalency is constant), then we would expect the band gap to increase with increasing ionicity, going for instance from Ge to GaAs and on to the II–VI semiconductor, zinc selenide (ZnSe). These general trends are indeed confirmed in fig. 2.16, where we plot the low temperature band gap (in electron volts) against the bond length for various group IV, III–V and II–VI compounds.

Tetrahedrally bonded III–V compounds span a very wide range of energy gaps, from 0.17 eV for InSb up to 6.2 eV in aluminium nitride (AlN). We note here a very useful relation between the band gap energy, E_g, in electron volts, and emission wavelength λ in microns, namely

$$\lambda E_g = 1.24 \, \mu m \, eV \tag{2.37}$$

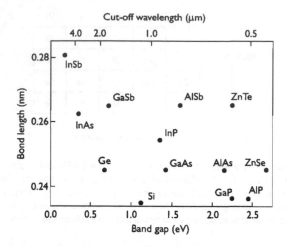

Figure 2.16 Plot of the low temperature energy gap (in eV) and the bond length (in Å) for various group IV, III–V, and II–VI compounds.

As light emission tends to occur due to transitions across the energy gap in semiconductors, we find that by varying the covalent and ionic contributions to bonding we can achieve light emission from bulk semiconductors at wavelengths ranging from the ultra-violet through to about 10 μm at the infra-red end of the spectrum.

References

Harrison, W. A. (1989) *Electronic Structure and the Properties of Solids: The Physics of the Chemical Bond*, Dover Publications, New York.

Harrison, W. A. (2000) *Applied Quantum Mechanics*, World Scientific Pub Co., Singapore.

Problems

2.1 Show that for two quantum wells of depth V_0 and width a separated by a barrier of width b, the energies of states of odd parity are found as solutions of eq. (2.10):

$$\left(\kappa^2 \coth\left(\frac{\kappa b}{2}\right) - k^2 \right) \sin(ka) + k\kappa \left(\coth\left(\frac{\kappa b}{2}\right) + 1 \right) \cos(ka) = 0$$

2.2 Show, by using eqs (1.53) and (1.54) that we can rewrite eq. (2.9) as

$$k \sin(ka/2) - \kappa \cos(ka/2) = \frac{\kappa^2 \sin(ka) + \kappa k \cos(ka)}{k \cos(ka/2) + \kappa \sin(ka/2)} \operatorname{sech}(\kappa b/2) e^{-\kappa b/2}$$

$$(2.38)$$

This reduces to eq. (1.50), the equation for even confined states in an isolated well, when the right-hand side equals zero. By re-arranging eq. (2.38) in the form

$$f(E) = \text{sech}(\kappa b/2) \exp(-\kappa b/2) \tag{2.39}$$

and then expanding $f(E)$ in a Taylor Series about E_0 (the isolated well ground state energy) show that $E_{gs}(b)$, the ground state energy in a coupled quantum well, varies for large barrier width b as $E_{gs}(b) = E_0 - Ce^{-\kappa b}$, where C is a constant which can in principle be determined from eqs (2.38) and (2.39).

2.3   Derive an equivalent expression to eq. (2.38) for the first excited state in a double quantum well, and hence show that the splitting between the ground and first excited state varies as $2Ce^{-\kappa b}$ for two weakly coupled square quantum wells.

2.4   The ground state energy level in a square well of width a and depth V_0, centred at the origin, is given by $\psi(x) = A\cos(kx)$, $|x| \leq a/2$, and $\psi(x) = De^{-\kappa|x|}$ for $|x| \geq a/2$, where k and κ have their usual meanings. By evaluating $\int_{-\infty}^{\infty} dx |\psi(x)|^2$, calculate the magnitude of the normalisation constants A and D in terms of k, κ and a. [This result can be useful when applying the variational method, as in the next question.]

2.5   Using the variational wavefunction $\psi(x) = \alpha\phi_L(x) + \beta\phi_R(x)$, where $\phi_L(x)$ and $\phi_R(x)$ are the isolated quantum well ground state wavefunctions defined in eq. (2.12), calculate each of the integrals I, II and IV in eq. (2.18) for the double square well potential. Hence show that the variational method also predicts that the splitting between the ground and first excited state energy varies as $2Ce^{-\kappa b}$ for two weakly coupled square quantum wells.

2.6   Show that the value of C calculated in problem 2.2 is the same as that calculated in problem (2.5)!!

2.7   We can write the wavefunctions for an s-state and for three p states on an isolated atom as $\phi_s(r) = f_s(r)$, $\phi_z(r) = f_p(r)\cos\theta$, $\phi_x(r) = f_p(r)\sin\theta\cos\phi$ and $\phi_y(r) = f_p(r)\sin\theta\sin\phi$, where (r, θ, ϕ) are spherical polar coordinates centred on the atomic nucleus, and $f_s(r)$ and $f_p(r)$ describe the radial variation of the s and p wavefunctions. Assuming that $f_s(r) \propto f_p(r)$, show that the hybrid orbital $\phi_h(r) = \frac{1}{2}[\phi_s(r) + \phi_x(r) + \phi_y(r) + \phi_z(r)]$ has maximum amplitude along the (111) direction ($\theta = \cos^{-1}(1/\sqrt{3}), \phi = \pi/4$).

2.8   As well as forming sp^3-bonded diamond crystals, carbon can also form sp^2-bonded graphite, where each carbon atom has three nearest neigbours, lying in the same plane as the carbon atom, with a bond angle of 120° between each pair of neighbours. Determine the form of the sp^2 hybrid orbitals on the central carbon atom if its three nearest neighbours all lie in the xy-plane, and one of the three neighbours is along the $+x$ direction.

2.9 N identical atoms each have a single electron at energy E_a. The atoms are brought together to form an N-membered ring, in which each atom interacts with its two neighbouring atoms, with an interaction of strength U ($U < 0$). It can be shown that the eigenstates, $\psi_n(r)$ of this ring can be expressed in the form

$$\psi_n(r) = \sum_{m=1}^{N} e^{i2\pi mn/N} \phi_m(r)$$

where $n = 0, 1, \ldots, N - 1$, and $\phi_m(r)$ is the atomic orbital on the mth atom. Show that the allowed energy levels in the N-membered ring are given by

$$E_n = E_a + 2U \cos(2\pi n/N).$$

Given that each energy level can contain two electrons, calculate the ground state binding energy per atom for all ring sizes between $N = 3$ and $N = 8$, and for $N = \infty$. The model here is appropriate to describe the interactions between neighbouring p_z orbitals in sp^2-bonded carbon. Hence, provide two reasons why 6-membered sp^2-bonded carbon rings are strongly favoured (e.g. as in benzene, C_6H_6) compared to other ring sizes.

2.10 Show that $\lambda E_g = 1.24\,\mu\mathrm{m}\,\mathrm{eV}$, where E_g is a photon energy in electron volts and λ is its wavelength in microns. The III–V alloy $InAs_xSb_{1-x}$ has a fraction x of the group V sites occupied by arsenic (As) atoms, and a fraction $(1 - x)$ occupied by antimony (Sb) atoms. The energy gap of $InAs_xSb_{1-x}$ (measured in eV) has been determined to vary with composition at room temperature as

$$E_g(x) = 0.17 + 0.19x + 0.58x(x - 1)$$

Determine the composition of the alloy with the lowest room temperature energy gap, and hence estimate an upper limit on the room temperature emission wavelength of conventional bulk III–V semiconductors.

Chapter 3

Band structure of solids

3.1 Introduction

We saw in the last chapter how we can build a good understanding of molecules and solids by describing the electronic structure using linear combinations of atomic orbitals. This method gives a very useful picture, particularly of trends in bonding properties. However, our earlier discussion gave at best a partial description of the electronic structure of solids. In particular, we only stated that isolated atomic and molecular energy levels broaden into bands of allowed energy states in solids, separated by forbidden energy gaps. In this chapter we consider in more detail the structure of these allowed energy bands.

There are about 10^{23} valence electrons which contribute to the bonding in each cubic centimetre of a typical solid. This implies that the calculation of the electronic structure should be a complex many-body problem, as the exact wavefunction and energy of each electron depend on those of all the others. However, there are at least two factors which considerably simplify the calculation of the energy spectrum.

First, it is found that in many cases each electron effectively sees a similar average potential as all the others, so that instead of having to solve something like a 10^{23} body problem, we can use an 'independent electron approximation', and calculate the energy spectrum using the one-electron Schrödinger equation introduced in Chapter 1. While we may not know the exact form of this average potential we expect that it should be closely related to the isolated atomic potentials of the atoms which form the solid.

Second, many interesting solid state materials are crystalline, with a periodic lattice. Because the ground state electronic structure must also be periodic, with the same charge distribution in each unit cell, we find that the potential $V(r)$ is periodic, with

$$V(r + R) = V(r) \tag{3.1}$$

where R is a vector joining the same point in two different unit cells, as illustrated in fig. 3.1. It can be shown that the individual electron

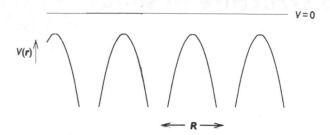

$V = 0$

$V(r)$

$\longleftarrow R \longrightarrow$

Figure 3.1 The variation in potential, $V(r)$, through a line of atoms in a periodic solid, with each atom separated by the vector R from its nearest neighbour in the line.

wavefunctions must reflect this periodicity, satisfying a condition referred to as Bloch's theorem.

We introduce Bloch's theorem in the next section, and describe how its application considerably simplifies the calculation and description of the electronic structure of crystalline solids. This is further illustrated in the following section where we extend the square well model of previous chapters to calculate the band structure of a one-dimensional (1-D) periodic array of square wells, using what is known as the Kronig–Penney (K–P) model.

There are several different techniques commonly used to calculate and develop an understanding of the electronic structure of solids. We provide an overview of three of these later in this chapter, using the K–P model to demonstrate their validity and applicability. We have already considered the tight-binding (TB) method in Chapter 2, based on isolated atom properties, and extend it in Section 3.4 to periodic solids. A very different approach is provided by the nearly free electron (NFE) model, described in Section 3.5. This starts from the assumption that the potential in a periodic solid is in many ways little different to that seen by an electron in free space, and calculates the band structure by treating the crystal potential as though it were only slightly perturbed from the constant, free space, potential. We show using the K–P model that, surprising as it may seem, there are situations where the two extremes, the TB and NFE models, each provide a good description of the electronic structure. This is for instance the case for tetrahedrally bonded semiconductors, as we illustrate in Section 3.6.

It is highly surprising that the NFE model should ever work in solids, as the electron–nucleus interaction has very sharp singularities, where the potential deviates very strongly from the flat, free space potential. We conclude this chapter by introducing in Section 3.7 the concept of a pseudopotential, showing how it is possible to modify the true potential to a much smoother 'pseudo' potential, which has the same calculated valence energy levels as the true potential and for which the NFE model then works remarkably well.

3.2 Bloch's theorem and band structure for a periodic solid

3.2.1 Bloch's theorem

We consider a solid with the periodic potential $V(r+R) = V(r)$, as defined in eq. (3.1). Bloch's theorem states that the wavefunctions of the one-electron Hamiltonian $H = -(\hbar^2/2m)\nabla^2 + V(r)$ can be chosen to have the form of a plane wave times a function with the periodicity of the lattice:

$$\psi_{nk}(r) = e^{ik \cdot r}u_{nk}(r) \tag{3.2a}$$

where

$$u_{nk}(r + R) = u_{nk}(r) \tag{3.3}$$

and where the subscript n refers to the nth state associated with the wavevector k. Combining eqs (3.2a) and (3.3) we can restate Bloch's theorem in an alternate (but equivalent) form

$$\psi_{nk}(r + R) = e^{ik \cdot R}\psi_{nk}(r) \tag{3.2b}$$

A full proof of Bloch's theorem can be found in several texts (e.g. Ashcroft and Mermin, Ibach and Lüth). We do not prove Bloch's theorem here but rather make its proof plausible by noting two consequences of eq. (3.2).

(1) *Periodic electron density*: We expect in a periodic solid that the electron probability density, $|\psi_{nk}(r)|^2$, can vary between different points within a given unit cell. This is allowed by eq. (3.2), as

$$|\psi_{nk}(r)|^2 = |e^{ik \cdot r}|^2|u_{nk}(r)|^2 = |u_{nk}(r)|^2 \tag{3.4}$$

and the function, $u_{nk}(r)$, although periodic, is not required to be constant, so can vary within a given unit cell. We also expect that the overall electron density should be equal at a given point r within one unit cell and the equivalent point $r + R$ within another unit cell. This also follows from Bloch's theorem, as from eq. (3.2b)

$$P_{nk}(r + R) = |\psi_{nk}(r + R)|^2 = |e^{ik \cdot R}|^2|\psi_{nk}(r)|^2 = P_{nk}(r) \tag{3.5}$$

so that there is equal probability of finding a given electron at r or at $r + R$, implying equal charge density at the two points.

(2) *Empty lattice model*: The wavefunctions for electrons in free space (where $V(r) \equiv 0$) can be chosen to take the form of plane waves, with the unnormalised wavefunction $\psi_k(r) = e^{ik\cdot r}$ describing a state with energy $E = \hbar^2 k^2/(2m)$. If we divide free space into a periodic array of identical boxes (giving what is referred to as the 'empty lattice'), then we can write each of the free space wavefunctions as the product of a plane wave times a constant (and therefore periodic) function:

$$\psi_k(r) = e^{ik\cdot r} \cdot 1 \tag{3.6}$$

Hence Bloch's theorem describes wavefunctions which reduce, as one would hope, to the correct form in the case where the periodic potential $V(r) \to 0$.

3.2.2 Electronic band structure

From Bloch's theorem, we can associate a wavevector k with each energy state E_{nk} of a periodic solid. It is often useful to plot a diagram of the energies E_{nk} as a function of the wavevector k, which is then referred to as the *band structure* of the given solid. Figure 3.2(a) shows the band structure for an electron in free space, which is described by the parabola $E = \hbar^2 k^2/(2m)$.

The free electron band structure is modified in several ways in a periodic solid. In particular, the wavevector k associated with a given energy state is no longer uniquely defined. This can be shown by considering a 1-D periodic structure, with unit cell of length L. We write the wavefunction for the nth state with wavenumber k as

$$\psi_{nk}(x) = e^{ikx} u_{nk}(x) \tag{3.7}$$

where e^{ikx} is a plane wave of wavenumber k, and $u_{nk}(x)$ is a periodic function, with $u_{nk}(x) = u_{nk}(x+L)$. To show that the wavenumber k is not uniquely defined, we can multiply eq. (3.7) by a plane wave with the periodicity of the lattice, $e^{i2\pi mx/L}$, and by its complex conjugate, $e^{-i2\pi mx/L}$, where m is an integer. This gives

$$\psi_{nk}(x) = e^{ikx} e^{i2\pi mx/L} e^{-i2\pi mx/L} u_{nk}(x)$$

$$= e^{i(k+2\pi m/L)x} \left(e^{-i2\pi mx/L} u_{nk}(x) \right) \tag{3.8}$$

where $e^{i(k+2\pi m/L)x}$ is a different plane wave to the original choice, and $e^{-i2\pi mx/L} u_{nk}(x)$ is still a periodic function, with period L. We refer to $G_m = 2\pi m/L$ as a *reciprocal lattice vector*, and note that the wavenumber k is then equivalent to the wavenumber $k + G_m$ in the given 1-D periodic structure.

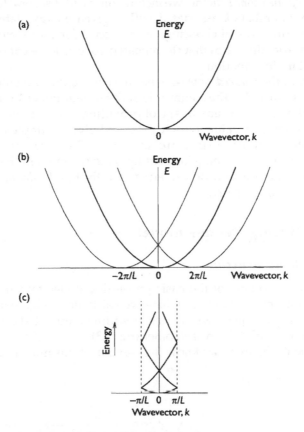

Figure 3.2 (a) The 'band structure' for a free electron, showing how the energy E varies quadratically with wavevector $\mathbf{k}$. (b) In a 1-D lattice with period L, the wavenumbers k and $k + 2\pi n/L$ are equivalent. In the *repeated zone* scheme, we then include in the band structure plot all wavenumbers $k + 2\pi n/L$ associated with each energy state. (c) In the *reduced zone* scheme, we choose the wavenumber for each energy state such that the magnitude of k is minimised. This then implies $-\pi/L < k \leq \pi/L$ for the 1-D lattice with period L.

If we now consider dividing 1-D free space into unit cells of length L, to create an empty lattice, we have several choices of how to plot the free electron band structure:

1   In the *extended zone* scheme, we try to associate a single, 'correct' wavenumber k with each state, as in fig. 3.2(a). While this is probably the best approach to take in the empty lattice, it becomes very difficult to assign a unique, 'correct' k to each state in a periodic crystal.

2   In the *repeated zone* scheme, we include on the plot several (in principle all) wavenumbers k associated with a given energy state. This gets over the difficulty of choosing the 'correct' k for each state, but it can be seen from fig. 3.2(b) that the repeated zone scheme contains a lot of redundant information.

3   Finally, in the *reduced zone* scheme, we choose the wavenumber k for each state such that the magnitude of the wavenumber k is minimised. This scheme has the advantage of providing a simple rule for assigning a preferred k value to each state, and gives a simple prescription for plotting the band structure, as in fig. 3.2(c). We will always use the reduced zone scheme for plotting band structure in this book. The reduced zone is also widely referred to as the first *Brillouin zone* for the given crystal structure.

3.3   The Kronig–Penney model

3.3.1   Full band structure

We can illustrate many of the basic properties of electrons in a periodic solid by using the K–P model, where we calculate the band structure of a periodic array of square wells, each of width a, separated by barriers of height V_0 and width b from each other (fig. 3.3).

From Bloch's theorem, we know the wavefunctions must be of the form

$$\psi_{nq}(x) = e^{iqx} u_{nq}(x) \tag{3.9a}$$

where

$$u_{nq}(x) = u_{nq}(x + a + b) \tag{3.9b}$$

and where we have chosen the letter q to symbolise the Bloch wavenumber. We first solve Schrödinger's equation within the first well to the right of the

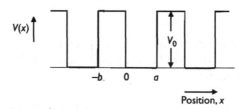

Figure 3.3 The K–P potential: a periodic array of square wells. We choose the wells here to be of width a, each separated by a barrier of height V_0 and width b from its immediate neighbours.

origin, where we choose to write the general solution for a state at energy E in the form

$$\psi(x) = A e^{ikx} + B e^{-ikx} \qquad 0 < x < a \qquad (3.10)$$

with $k^2 = 2mE/\hbar^2$, as before. The general solution to Schrödinger's equation in the first barrier to the left of the origin is given by

$$\psi(x) = C e^{\kappa x} + D e^{-\kappa x} \qquad -b < x < 0 \qquad (3.11)$$

where $\kappa^2 = 2m(V_0 - E)/\hbar^2$, and we assume $E < V_0$.

We now introduce four boundary conditions to determine the allowed solutions of Schrödinger's equation. The first two are easily chosen, requiring that the wavefunction ψ and its derivative $d\psi/dx$ are continuous at the well/barrier interface at $x = 0$:

$$\psi(0): \quad A + B = C + D \qquad (3.12a)$$

$$\psi'(0): \quad ikA - ikB = \kappa C - \kappa D \qquad (3.12b)$$

The two remaining boundary conditions are derived using Bloch's theorem. We apply eq. (3.2b) at $x = a$, so that $\psi_w(a) = \psi_b(-b)e^{iq(a+b)}$ and likewise for the derivatives $\psi'_w(a) = \psi'_b(-b)e^{iq(a+b)}$:

$$\psi_w(a): \quad A e^{ika} + B e^{-ika} = (C e^{-\kappa b} + D e^{\kappa b})e^{iq(a+b)} \qquad (3.12c)$$

$$\psi'_w(a): \quad ikA e^{ika} - ikB e^{-ika} = (\kappa C e^{-\kappa b} - \kappa D e^{\kappa b})e^{iq(a+b)} \qquad (3.12d)$$

We can then derive the condition for allowed energy levels by solving the 4×4 determinant involving the four unknowns A, B, C, and D, as was done with eq. (1.52) in Chapter 1. The energy levels associated with Bloch wavenumber q are then found to be solutions of the equation

$$\cos(q(a+b)) = \cos(ka)\cosh(\kappa b) + \frac{1}{2}\left(\frac{\kappa}{k} - \frac{k}{\kappa}\right)\sin(ka)\sinh(\kappa b) \qquad (3.13)$$

We can deduce one particularly important result from this equation, namely, that there will always be energy gaps between the allowed energy bands in the K–P model. To see this, we first note that the left-hand side of eq. (3.13), $\cos(q(a+b))$, can only take values between -1 and 1. If we choose $ka = n\pi$, then the right-hand side reduces to $\pm\cosh(\kappa b)$, as $\sin(n\pi) = 0$ and $\cos(n\pi) = \pm 1$. The magnitude of $\cosh(\kappa b)$ is always greater than 1 for states which are bound within the well. Hence, there can be no solutions to eq. (3.13) for $k = n\pi/a$, implying no solutions are possible with energy $E = (\hbar^2/2m)(n\pi/a)^2$, and likewise for neighbouring values of the energy. This is illustrated in fig. 3.4(a), where we plot an example of the variation

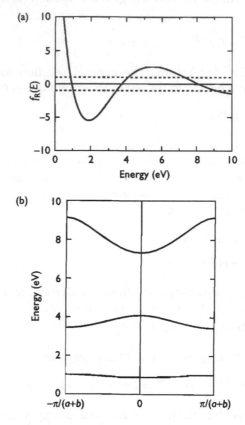

Figure 3.4 (a) The horizontal dashed lines at +1 and at −1 show the upper and lower limits of the left-hand side of eq. (3.13). The smooth curve shows how the right-hand side of eq. (3.13), $f_R(E)$, varies with energy E for the case where $a = 5\,Å$, $b = 1.5\,Å$ and $V_0 = 10\,eV$. It can be seen that there are only certain bands of energy where there can be allowed solutions of eq. (3.13) (around 1, 4, and 8 eV in this case), separated from each other by energy gaps where there are no allowed solutions. (b) shows the band structure in the K–P model, with energy plotted against wavenumber q for the case considered in (a) ($a = 5\,Å$, $b = 1.5\,Å$ and $V_0 = 10\,eV$). It can be seen that the band edges occur at $q = 0$ or $\pi/(a + b)$, that is, where the left-hand side of eq. (3.13) equals ±1.

with energy, E, of the right-hand side of eq. (3.14), $f_R(E)$, superimposed on a graph of the range of allowed values of the left-hand side, showing clearly the existence of allowed and forbidden energy regions.

Figure 3.4(b) shows the calculated band structure for the K–P model, with E plotted against q in the first Brillouin zone (for which $|q| \leq \pi/(a + b)$), for the case with well width $a = 5\,\text{Å}$, barrier width, $b = 1.5\,\text{Å}$, and barrier height $V_0 = 10\,\text{eV}$. The energies vary continuously with q within each band, and the band edges always occur at high symmetry points: either at the centre of the Brillouin zone ($q = 0$), or else at the Brillouin zone edge ($q = \pi/(a+b)$). Although derived using the K–P model, this result is general for 1-D periodic structures, where the band extrema and band gaps are always associated with these two high symmetry points. We will see below that in two and three dimensions band extrema may in addition occur also at lower symmetry points, due to band mixing and anti-crossing effects (see fig. 3.14).

We would hope with eq. (3.13) that as $\kappa b \to \infty$, the allowed energy levels should approach those for an isolated finite quantum well. Equation (3.13) can be rewritten as

$$\cos(ka) + \frac{1}{2}\left(\frac{\kappa}{k} - \frac{k}{\kappa}\right)\sin(ka)\tanh(\kappa b) = \frac{\cos(q(a+b))}{\cosh(\kappa b)} \tag{3.14}$$

As $\kappa b \to \infty$, $\tanh(\kappa b) \to 1$, while $\cosh(\kappa b) \to \infty$, so that eq. (3.14) then reduces to

$$\cos(ka) + \frac{1}{2}\left(\frac{\kappa}{k} - \frac{k}{\kappa}\right)\sin(ka) = 0 \tag{3.15}$$

But we saw in Chapter 1 (eq. (1.54)) that this is just the condition to determine the energy levels for an isolated well. Hence, as for the double quantum well in Chapter 2, we expect that we should be able to use the TB (linear combination of atomic orbitals) method to determine the energy levels in crystalline solids. We also expect that only the higher energy (valence) levels will contribute to the bonding, while the deeper (core) levels will be largely unperturbed by the neighbouring atoms.

3.3.2 High symmetry energy states

We saw in Chapter 1 how we can use symmetry arguments to simplify the calculation of the confined state energies in an isolated square well, with the energy levels for even states given by eq. (1.50):

$$k\tan(ka/2) = \kappa \tag{3.16a}$$

and for odd states by eq. (1.51)

$$k\cot(ka/2) = -\kappa \tag{3.16b}$$

We can also use symmetry in the K–P model to simplify the calculation of the confined state energies for the Bloch wavevectors $q = 0$ and

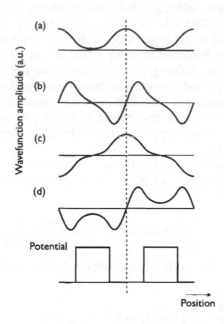

Figure 3.5 The wavefunctions of the zone centre ($q = 0$) states in the K–P model are either (a) even about the centre of each well and about the centre of each barrier or (b) odd about the centre of each well and barrier. At the zone edge ($q = \pi/(a + b)$) the wavefunctions can be either (c) even about the well centre and odd about the barrier centre or (d) odd about the well centre and even about the barrier centre.

$q = \pi/(a + b)$. It can be shown that the states at $q = 0$ are either symmetric about the centre of the well and the centre of the barrier (as in fig. 3.5(a)) or else anti-symmetric about the centre of the well within one unit cell, while having the same value at x and at $x + (a + b)$ (fig. 3.5(b)). The allowed symmetries of the states at $q = \pi/(a + b)$ are illustrated in fig. 3.5(c) and (d). Because of the symmetry, we can write the wavefunction $\psi_w(x)$ within the well region in fig. 3.5(a) as

$$\psi_w(x) = A\cos(k(x - a/2)) \qquad 0 < x < a \qquad (3.17a)$$

where $\psi_w(x)$ is chosen to be symmetric about $x = a/2$. Likewise the wavefunction $\psi_b(x)$ within the barrier to the left of the origin can be chosen to be symmetric about $x = -b/2$:

$$\psi_b(x) = C\cosh(\kappa(x + b/2)) \qquad -b < x < 0 \qquad (3.17b)$$

If we then require the wavefunction and its derivative to be continuous at the origin, we find

$$\psi(0): \quad A\cos(ka/2) = C\cosh(\kappa b/2) \tag{3.18a}$$

$$\psi'(0): \quad kA\sin(ka/2) = \kappa C\sinh(\kappa b/2) \tag{3.18b}$$

Dividing (3.18b) by (3.18a) we obtain for the states in fig. 3.5(a) that

$$k\tan(ka/2) = \kappa\tanh(\kappa b/2) \tag{3.19a}$$

A similar analysis gives, respectively, for the states in fig. 3.5(b)–(d)

$$k\cot(ka/2) = -\kappa\coth(\kappa b/2) \tag{3.19b}$$

$$k\tan(ka/2) = \kappa\coth(\kappa b/2) \tag{3.19c}$$

$$k\cot(ka/2) = -\kappa\tanh(\kappa b/2) \tag{3.19d}$$

The results of eq. (3.19) are particularly elegant. They illustrate clearly how the band edge energies in the solid evolve both from the isolated well values ($b = \infty$) and at the opposite extreme from the 'empty lattice' results ($b = 0$).

3.4 The tight-binding method

We outline in this section how the TB method can be used to successfully calculate the band structure of the K–P model from 'first principles'. The calculation provides an excellent description of the energy spectrum for bound states up to relatively small interwell separations, b, and also illustrates several general features of the TB method. We will see how the magnitude of the Hamiltonian matrix elements linking the atomic levels in neighbouring quantum wells decreases both with increasing well separation, and also as a state becomes more tightly bound within a given well. We shall also see that the TB method works best for the lowest lying energy levels, becoming less acccurate for higher-lying excited states. This again is a general feature of the method.

We consider the periodic array of square wells, illustrated in fig. (3.3), with the nth well defined in the region $nL < x < nL + a$, where $L = a + b$. We first solve Schrödinger's equation to find the energy levels E_m and normalised wavefunctions, $\phi_m(x)$ for an isolated quantum well defined between $0 < x < a$. We presume that the states in the mth energy band of the K–P model can be formed using a linear combination of the mth energy states in each of the wells. In order to satisfy Bloch's theorem (eq. 3.2), the wavefunction, $\psi_{mq}(x)$, for the state in the mth band with wavenumber q is then given by

$$\psi_{mq}(x) = \sum_{n=-\infty}^{\infty} e^{inqL} \phi_m(x - nL) \tag{3.20}$$

where $\phi_m(x - nL)$ is the 'atomic orbital' associated with the mth state in the nth quantum well. Schrödinger's equation is then given by

$$H\psi_{mq}(x) = E_{mq}\psi_{mq}(x) \tag{3.21}$$

We evaluate E_{mq} using a similar technique to that used in Section 2.3 for the H_2 molecule. We multiply both sides of eq. (3.21) from the left-hand side by $\phi_m^*(x)$, the mth state in the zeroth well, and then integrate with respect to x, to find

$$\sum_{n=-\infty}^{\infty} e^{inqL} \int_{-\infty}^{\infty} dx\, \phi_m^*(x) H\phi_m(x - nL)$$

$$= E_{mq} \sum_{n=-\infty}^{\infty} e^{inqL} \int_{-\infty}^{\infty} dx\, \phi_m^*(x)\phi_m(x - nL) \tag{3.22a}$$

To make this and subsequent equations more manageable, we introduce a compact notation at this point, rewriting eq. (3.22a) using Dirac notation (see Appendix D) as

$$\sum_{n=-\infty}^{\infty} e^{inqL} \langle \phi_{m0}|H|\phi_{mn}\rangle = E_{mq} \sum_{n=-\infty}^{\infty} e^{inqL} \langle \phi_{m0}|\phi_{mn}\rangle \tag{3.22b}$$

where $|\phi_{mn}\rangle = \phi_m(x - nL)$. Because each of the basis functions $|\phi_{mn}\rangle$ decays exponentially outside its own well, the overlap integrals $\langle\phi_{m0}|H|\phi_{mn}\rangle$ and $\langle\phi_{m0}|\phi_{mn}\rangle$ decrease very rapidly with increasing n, so we need only retain the on-site terms ($n = 0$), and the two nearest neighbour terms ($n = 1$ and -1) in eq. (3.22) (see fig. 3.6). We can also show from symmetry that $\langle\phi_{m0}|H|\phi_{m1}\rangle = \langle\phi_{m0}|H|\phi_{m,-1}\rangle$, and that $\langle\phi_{m0}|\phi_{m1}\rangle = \langle\phi_{m0}|\phi_{m,-1}\rangle$. As $e^{iqL} + e^{-iqL} = 2\cos(qL)$, eq. (3.22) then reduces to

$$\langle\phi_{m0}|H|\phi_{m0}\rangle + 2\langle\phi_{m0}|H|\phi_{m1}\rangle \cos(qL) = E_{mq}\{1 + 2S_m \cos(qL)\} \tag{3.23}$$

where we have defined $S_m = \langle\phi_{m0}|\phi_{m1}\rangle$ as the overlap integral between the mth basis functions centred on well zero and on well 1.

We can in principle calculate exactly the overlap interaction between neighbouring sites, $\langle\phi_{m0}|H|\phi_{m1}\rangle$ on the left-hand side of eq. (3.23). We introduce the interaction parameter, V_m, as in Chapter 2, such that

$$\langle\phi_{m0}|H|\phi_{m1}\rangle = \int d^3r\, \phi_{m0}^*(x)H\phi_{m1}(x) = V_m \tag{3.24}$$

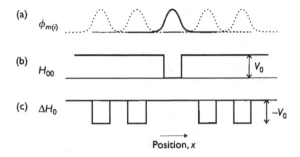

(a) $\phi_{m(i)}$

(b) H_{00} V_0

(c) ΔH_0 $-V_0$

Position, x

Figure 3.6 (a) In the TB method, we write each wavefunction as a linear combination of isolated quantum well wavefunctions. Because these basis functions decay exponentially with distance we generally need only consider the on-site and nearest neighbour interactions. The thick solid line shows the basis function associated with the 0th site. For the TB calculations, we can express the total K–P potential of fig. 3.3 as the sum of (b) the potential H_{00} due to an isolated quantum well, plus (c) ΔH_0, the difference in potential between the full K–P potential and the isolated well potential, H_{00}.

where V_m is a measure of the strength of the interaction between atomic orbitals centred on neighbouring wells, with the magnitude of V_m increasing exponentially with decreasing separation between the wells.

$\langle \phi_{m0}|H|\phi_{m0}\rangle$ is evaluated in a similar manner to that used in Chapter 2. We first write $H = H_{0p} + \Delta H_p$, where H_{0p} is the isolated square well potential associated with the pth well ($p = 0$ or 1), and ΔH_p is the difference in potential between the full K–P model and the isolated pth well (fig. 3.6(c)). Then

$$H\phi_m(x - pL) = H_{0p}\phi_m(x - pL) + \Delta H_{0p}\phi_m(x - pL)$$
$$= E_m\phi_m(x - pL) + \Delta H_p\phi_m(x - pL) \qquad (3.25)$$

Substituting eq. (3.25) into $\langle \phi_{m0}|H|\phi_{m0}\rangle$ we find

$$\langle \phi_{m0}|H|\phi_{m0}\rangle = E_m\langle \phi_{m0}|\phi_{m0}\rangle + \langle \phi_{m0}|\Delta H_0|\phi_{m0}\rangle$$
$$= E_m + \langle \phi_{m0}|\Delta H_0|\phi_{m0}\rangle$$
$$= E_m + \Delta E_m \qquad (3.26)$$

where $\Delta E_m = \langle \phi_{m0}|\Delta H_0|\phi_{m0}\rangle$ is often defined as the self-energy shift of the given level, and reflects the change in energy of the orbital $\phi_m(x)$ on going from an isolated well to the crystalline potential (fig. 3.6). Substituting eqs (3.24) and (3.26) back in eq. (3.23) we then calculate E_{mq}, the energy of

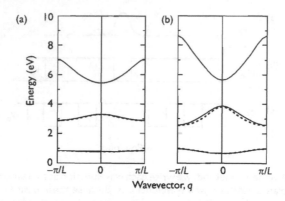

Figure 3.7 Comparison of the exact band structure for the K–P potential (solid lines)
and the band structure calculated using the TB method (dashed lines).
(a) Widely separated wells ($a = 5\,\text{Å}$, $V_0 = 5\,\text{eV}$, $b = 3\,\text{Å}$), where agreement is excellent. Note that the TB method is used here to fit the two
lowest bands only, as the next band cannot be formed as a linear combination of states confined in an isolated quantum well. (b) Moderate barrier
widths ($b = 1.5\,\text{Å}$), where the agreement is still very good, although it can
be seen that the two sets of results are beginning to diverge for the second
band.

the state with wavevector q in the mth band as

$$E_{mq} = \frac{E_m + \Delta E_m + 2V_m \cos(qL)}{1 + 2S_m \cos(qL)} \qquad (3.27a)$$

$$= E_m + 2V_m \cos(qL) \qquad (3.27b)$$

where we assume that the overlap term S_m and the self-energy shift ΔE_m
are sufficiently small that they can be neglected.

Figure 3.7 compares the exact K–P and the TB band structure calculated using eq. (3.27a) for two particular cases, first, where the wells are
widely separated (large b) and second, when the wells have been brought
closer together (moderate b). It is clear that the TB band structure (dashed
lines) is in excellent agreement with the exact solution (solid lines) for the
large b case. The agreement still remains very good in the second case, for
moderate b, particularly for the lowest band.

The agreement in fig. 3.7(b) could be further improved by 'parameteris-
ing' the model, that is, choosing TB interaction parameters V_m and orbital
self-energy shifts, ΔE_m such that the TB band structure is then fitted to be in
good agreement with the exact bands. This is what generally occurs in practical applications in solid state physics, where neither the basis functions
nor interaction parameters are calculated exactly but are instead found

by fitting to experimentally determined band structures. This approach is justified because the trends in the fitted parameters generally do show the expected behaviour. The TB method has been widely used to successfully analyse trends in the band structure and electronic properties of a broad range of solids [see for instance the texts by Harrison and by Pettifor].

We have demonstrated the principle of the TB method here by considering an example where a given orbital on a particular site had a significant interaction with only one orbital on each neighbouring site. More orbitals must generally be included for practical applications: we saw for instance in Chapter 2 how at least one s and three p orbitals need to be included per site to describe the electronic properties of tetrahedrally bonded semiconductors. The extension of the TB method to include several orbitals per site is in principle straightforward. Further details can be found, for instance, in Harrison.

3.5   The nearly free electron method

We can also view an infinite set of quantum wells from the opposite extreme. Figure 3.8 shows how the potential approaches the free space potential, $V = 0$ everywhere, as the interwell separation $b \to 0$. We might expect for such narrow barriers that we could treat the energy states using the 'nearly free electron' (NFE) model, where we build up the wavefunctions using linear combinations of free electron energy states.

We consider the infinite set of square wells shown in fig. 3.8, where again the period $L = a + b$, but b is now small. The solid lines in fig. 3.9 show the free space band structure for the 'empty lattice' with period L. We can associate an integer n with each free electron state at wavevector q, with the energy of the nth state given by

$$E_{nq} = \frac{\hbar^2}{2m} \left(q + \frac{2\pi n}{L} \right)^2 \tag{3.28}$$

Figure 3.8 The K–P potential for a very narrow barrier width, b. As $b \to 0$, the K–P potential approaches the free space potential ($V = 0$ everywhere).

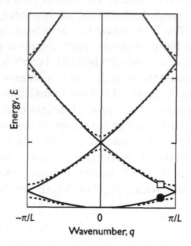

Figure 3.9 The solid curves show the free-electron band structure for the 'empty lattice' with period L. In the NFE model, we assume that the wavefunctions
of, for example, the two lowest states at a value of q near π/L can be
approximated by a linear combination of the two free electron wavefunctions (● and □) at that q-value. The dotted lines illustrate the resulting
band structure in the NFE model.

and the wavefunction by

$$\psi^0_{nq} = \frac{1}{\sqrt{L}}e^{i(q+(2\pi n/L))x} = \frac{1}{\sqrt{L}}e^{iqx}e^{i2\pi nx/L} \tag{3.29}$$

where we have chosen to normalise the wavefunction within each unit cell,
and we have also deliberately separated each wavefunction into the plane
wave $\exp(iqx)$ times a cell-periodic function, $\phi_n(x) = L^{-1/2} \exp(i2\pi nx/L)$.
The set of periodic functions $\phi_n(x)$ form a complete set, as defined in
Section 1.7, so that any periodic function $f(x)$ with period L can be expanded
in terms of the complete set $\phi_n(x)$:

$$f(x) = \sum_{n=-\infty}^{\infty} a_n\phi_n(x) \tag{3.30}$$

Returning to the periodic potential, $V(x)$, of fig. 3.8, where

$$V(x) = V_0 \qquad \frac{-b}{2} < x < \frac{b}{2}$$

$$V(x) = 0 \qquad \frac{b}{2} < x < a + \frac{b}{2} \tag{3.31}$$

the 'NFE' method assumes that the exact wavefunctions ψ_{nq} for this potential can be written efficiently as linear combinations of the free electron

wavefunctions

$$\psi_{nq}(x) = \frac{1}{\sqrt{L}} e^{iqx} \sum_{m=-\infty}^{\infty} a_{mnq} e^{i2\pi mx/L} \tag{3.32}$$

In calling this an 'efficient' expansion method, we mean that most of the coefficients a_{mnq} will be of order zero, and only free electron states close in energy to the given state will contribute significantly to its wavefunction.

We illustrate this using the two lowest states at a wavevector q close to the Brillouin zone edge at π/L. We presume that the NFE wavefunctions for these two states can be made up as a linear combination of the two free electron wavefunctions, with wavevectors q and $q - 2\pi/L$ (labelled by $\bullet$ and $\square$, respectively in fig. 3.9):

$$\psi_q(x) = \alpha/L^{1/2} e^{iqx} + \beta/L^{1/2} e^{i(q-2\pi/L)x} \tag{3.33}$$

We wish to find the values of the coefficients α and β to satisfy Schrödinger's equation

$$H\psi_q(x) = H(\alpha\, e^{iqx} + \beta\, e^{i(q-2\pi/L)x})/L^{1/2}$$

$$= E(\alpha\, e^{iqx} + \beta\, e^{i(q-2\pi/L)x})/L^{1/2} \tag{3.34}$$

We can write

$$H = H_0 + V(x) \tag{3.35}$$

where H_0 is the free-space Hamiltonian and $V(x)$ the change in the free-space potential due to the barriers, given by eq. (3.31). We can solve the Hamiltonian equation (3.34) using exactly the same approach as was used for the LiH molecule in Section 2.4. If we multiply eq. (3.34) from the left by $e^{-iqx}/L^{1/2}$ and integrate over a unit cell between $-b/2$ and $a+b/2$, we get the first of two linear equations which must be solved:

$$\frac{1}{L} \int_{-b/2}^{a+b/2} dx\, e^{-iqx} (H_0 + V(x))(\alpha\, e^{iqx} + \beta\, e^{i(q-2\pi/L)x})$$

$$= \frac{E}{L} \int_{-b/2}^{a+b/2} dx\, e^{-iqx}(\alpha\, e^{iqx} + \beta\, e^{i(q-2\pi/L)x}) \tag{3.36}$$

Using the orthonormality properties of the two wavefunctions, this equation reduces to

$$\alpha \left(\frac{\hbar^2 q^2}{2m} + \frac{b}{L} V_0 \right) + \beta \frac{V_0}{\pi} \sin\left(\frac{\pi b}{L} \right) = \alpha E \tag{3.37a}$$

while pre-multiplying by $e^{-i(q-2\pi/L)x}$ gives as the second linear equation

$$\alpha \frac{V_0}{\pi} \sin\left(\frac{\pi b}{L}\right) + \beta \left[\frac{\hbar^2 (q - 2\pi/L)^2}{2m} + \frac{b}{L} V_0 \right] = \beta E \tag{3.37b}$$

The two lowest energy levels at wavevector q close to π/L are then calculated by solving eq. (3.37), which gives

$$E = \frac{b}{L} V_0 + \frac{\hbar^2}{2m} \left[\left(\frac{\pi}{L}\right)^2 + \left(q - \frac{\pi}{L}\right)^2 \right] \pm \sqrt{ \frac{\hbar^4 \pi^2}{m^2 L^2} \left(\frac{\pi}{L} - q\right)^2 + \frac{V_0^2}{\pi^2} \sin^2 \frac{\pi b}{L} } \tag{3.38}$$

as illustrated by the dotted lines in fig. 3.9. Hence, the introduction of the periodic potential $V(x)$ of eq. (3.31) introduces an energy gap at the Brillouin zone boundary, at $q = \pi/L$, with the energy gap, E_g, between the two lowest lying states calculated by setting $q = \pi/L$ in eq. (3.38), in which case

$$E_g = \frac{2V_0}{\pi} \sin\left(\frac{\pi b}{L}\right) \tag{3.39a}$$

It can be shown that the nth gap at $q = \pm\pi/L$ is given by

$$E_g = \frac{2V_0}{(2n-1)\pi} \sin\left(\frac{(2n-1)\pi b}{L}\right) \tag{3.39b}$$

while the nth gap at $q = 0$ is given by

$$E_g = \frac{V_0}{n\pi \sin(2n\pi b/L)} \tag{3.39c}$$

    Figure 3.10 compares the exact band structure and the NFE band structure for two particular cases, first for a very thin barrier (b close to zero) and second when the wells have been moved further apart to the same intermediate value of b used in fig. 3.7(b) for the TB method. The NFE band structure (dotted lines) was calculated by assuming that only two free electron states contribute to the NFE wavefunction for each state. The NFE results are in excellent agreement with the exact solution (solid lines) for the small b case and, interestingly, remain in surprisingly good agreement, at least for the higher levels, in the second, intermediate b case.

    We thus have the apparent paradox that at intermediate well separations we can get a good description of the true band structure either by assuming that the wavefunctions are NFE-like, or by presuming that they can be built up using linear combinations of tightly bound atomic orbitals. The fact that

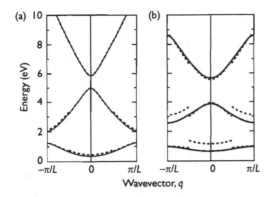

Figure 3.10 Comparison of the exact band structure for the K–P potential (solid
lines) and the band structure calculated using the simplest form of the
NFE model (dotted lines). (a) Very thin barrier ($a = 5\,\text{Å}$, $V_0 = 5\,\text{eV}$,
$b = 0.5\,\text{Å}$), where agreement is excellent, and (b) for moderate barrier
width ($b = 1.5\,\text{Å}$), where the agreement is poor for the lowest band,
but still good for the higher bands. In each case, states with $|q| < \pi/2L$
were calculated using zone centre free-electron states, while states with
$\pi/2L < |q| < \pi/L$ were calculated using zone edge free-electron states.

both models work so well at the same time is a testament to the effectiveness
of the variational method. Figure 3.11 compares the exact wavefunction
(solid line) for the lowest state at $q = \pi/L$ in the moderate barrier case with
the wavefunctions used in the NFE and TB methods (dotted and dashed
lines respectively). Both trial functions show good agreement with the exact
function, and we have seen in Chapter 1 that the accuracy of the calculated
energy levels is generally better than that of the assumed wavefunctions
used in the variational method.

Figure 3.12 examines further the range of validity of the NFE and TB
methods. The solid lines indicate the calculated variation of the band edge
energies as a function of barrier width b for the case of a well of width
$a = 5\,\text{Å}$ and depth $V_0 = 5\,\text{eV}$. The dashed lines show the calculated energies using the TB model, where the band edge energies are assumed from
eq. (3.27) to vary as

$$E = \frac{E_m + \Delta E_m \pm 2V_m}{1 \pm 2S_m} \tag{3.27}$$

while the dotted lines show the calculated NFE results, using eq. (3.39).
It is quite remarkable to realise that the accuracy of both models could
be further improved by including interactions between more basis states
in the energy level calculations, so that there can clearly be a significant

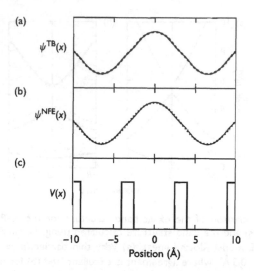

Figure 3.11 Comparison of the lowest K–P wavefunction at $q = \pi/L$ (solid line) with the wavefunctions used (a) in the TB (dashed) and (b) NFE (dotted) models, for the case of a moderate barrier width ($b = 1.5\,\text{Å}$, $a = 5\,\text{Å}$, $V_0 = 5\,\text{eV}$), for which case both the TB and NFE methods give good agreement with the exact result.

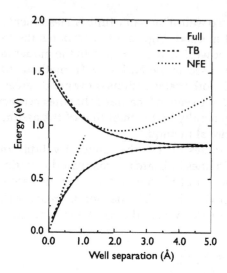

Figure 3.12 Calculated variation of the band edge energies as a function of barrier width, b, for the K–P potential with well width $a = 5\,\text{Å}$ and barrier height $V_0 = 5\,\text{eV}$. The solid lines show the exact results while the dashed and dotted lines show the energies calculated using the TB and NFE methods, respectively.

range of well separations where the K–P model can be solved using either the TB or NFE model. The same is also true of the electronic properties of many materials. As a general rule, many simple metals such as sodium or aluminium are best described by the NFE model. The properties of insulators such as silica (SiO_2) and small molecules are often most easily understood using the TB method. There is also, however, a wide range of materials which can be described equally effectively using the NFE or TB method, as we illustrate below for the semiconductor, germanium.

3.6   Band structure of tetrahedral semiconductors

We saw in Chapter 2 how the crystal structure of silicon or germanium can be described in terms of a face-centred-cubic (FCC) lattice, with two atoms in each unit cell of the lattice. The unit cell is the basic building block of the lattice: we can construct an FCC lattice by fitting together a set of blocks shaped as in fig. 3.13(a). For silicon or germanium, each of these blocks contains two atoms, one located say at the origin and the other at ($a/4$, $a/4$, $a/4$), where a is the length of the side of the cube in the FCC lattice. It can be shown that the reciprocal lattice of an FCC lattice is a body-centred-cubic (BCC) lattice, with the first Brillouin zone for the reciprocal lattice then given by a truncated octahedron (fig. 3.13(b)). Because the FCC lattice is widespread and of such importance, several high symmetry points and lines in its first Brillouin zone have received specific names. The Brillouin zone centre, where the wavevector $k = 0$, is referred to as the Γ point, while the zone edge along the [001] and related directions (at $k = (0, 0, \pi/2a)$) is the X point, with the zone edge along [111] (at $k = (\pi/4a, \pi/4a, \pi/4a)$) called the L point. The line of k points joining Γ and X are referred to as the Δ direction, and the line from Γ to L as the Λ direction.

We cannot plot the band structure for every wavevector k in the first Brillouin zone of a 3-D solid, as it is impossible to project every point in three dimensions onto a 2-D plot. What do we do instead? Usually we plot the band structure along a number of high symmetry lines, such as the Δ direction from Γ to X (associated with waves propagating along the cubic axes directions) or the Λ direction from Γ to L (associated with waves propagating along body-diagonal directions such as along the $(1, 1, 1)$ direction in the cube).

The band structure of germanium is illustrated in fig. 3.14(a) and (b), calculated in (a) using the TB method, with four atomic orbitals per site, and in (b) using a NFE-type calculation. Both band structures were obtained empirically, choosing interaction parameters to enable a good fit to experimental data. There are eight valence electrons in each unit cell of germanium. Each band in fig. 3.14 can take two electrons from each unit cell (one with 'up' spin and one with 'down' spin). The four lowest

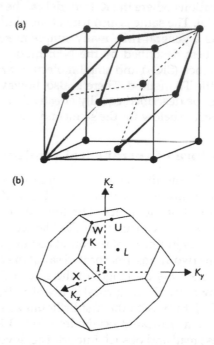

(a)

(b)

Figure 3.13 (a) The unit cell (basic building block) of a FCC lattice: the FCC lattice can be constructed by filling all space with a set of blocks identical to that shown. (b) The first Brillouin zone for the reciprocal lattice of an FCC lattice. Several high symmetry points in this Brillouin zone have been given specific names, some of which are indicated in the figure. The shape shown is the unit cell for a BCC lattice, because the reciprocal lattice of an FCC lattice is a BCC lattice. (From H. P. Myers (1997) *Introductory Solid State Physics*, 2nd edn.)

bands then form the filled valence (bonding) bands, and are separated by the energy gap, E_g, from the empty conduction bands. It can be seen that the two calculations are in very good agreement with each other for the valence bands, confirming the validity of either approach, although clearly the agreement becomes less good at higher energy, in the conduction bands. This problem is related to the use of only a small number of basis states in the TB model, and can be overcome, at least in part, by using more basis states and including extra interactions in the Hamiltonian.

Figure 3.14(c) gives the free-electron band structure for an empty germanium lattice. Each free-electron state has been shifted into the first Brillouin zone, as was done earlier in fig. 3.2(c). It is clear that there are many similarities between the true bands of fig. 3.14(b) and the free-electron bands, further justifying and motivating the use of NFE-type models.

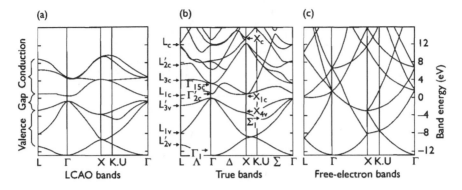

Figure 3.14 The energy band structure of germanium, calculated (a) using the TB method, and (b) using a NFE-type method. It can be seen that there is very good agreement between the four lowest bands in the two cases (at energy $E < 0$ eV). Part (c) shows for comparison the free-electron band structure for an empty germanium lattice. (Reproduced from Harrison (1989) with permission.)

3.7   The use of 'pseudo' potentials

The true potential in any solid varies very rapidly near each atomic nucleus, becoming strongly attractive due to the Coulomb interaction between the nucleus and electrons. Because the electron total energy must be conserved at every point, the kinetic energy then becomes very large near the nuclei, implying rapid oscillation of the valence wavefunctions, as illustrated in fig. 3.15. It is, therefore, very surprising that the band structure of any solid could ever be calculated using a small number of low energy plane waves – how can such plane waves reproduce the true wavefunction near a nucleus?

The answer is they cannot. We could, however, have provided an alternative argument as to why the wavefunctions oscillate so rapidly near the atomic nuclei. Most atoms have tightly bound core energy levels, ε_{ci}, with their associated atomic orbital wavefunctions, $\phi_{ci}(r)$, as discussed in Appendix B. We know that for any potential the solutions of Schrödinger's equation must be orthogonal to each other. The valence states, which dominate the solid's electronic properties, must then be orthogonal to the core states. We can therefore construct the valence states from orthogonalised plane waves (the OPW method), given by

$$\phi_v^k(r) = e^{ik \cdot r} + \sum_i a_{ik}\phi_{ci}(r) \tag{3.40}$$

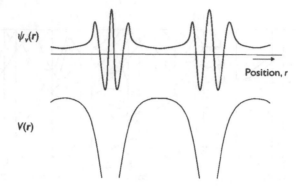

Figure 3.15 The valence wavefunctions in a solid are expected to oscillate rapidly near each nucleus, because of the strong Coulomb interaction between the nucleus and the electron. How then can the valence wavefunction be approximated by a small number of plane waves when we use the NFE model?

where the coefficients a_{ik} are chosen to ensure the valence basis functions $\phi_v^k(r)$ are orthogonal to the core states within each unit cell, that is,

$$\int d^3r\, \phi_{ci}^*(r)\phi_v^k(r) = 0 \tag{3.41}$$

This can be achieved by choosing the a_{ik} values such that

$$a_{ik} = -\int d^3r\, \phi_{ci}^*(r)e^{ik\cdot r} \tag{3.42}$$

The orthogonalised plane waves of eq. (3.40) are, therefore, constructed to have the necessary rapid oscillations in the core regions while having a slow plane-wave-like variation in the remainder of the structure. It is then possible to calculate the band structure using a relatively small number of OPW basis functions.

We do this by using linear combinations of orthogonalised plane waves to solve the full crystal Hamiltonian for the valence states

$$H \sum_m \alpha_{mn}\phi_v^{km}(r) = E_{kn} \sum_m \alpha_{mn}\phi_v^{km}(r) \tag{3.43}$$

where the $\phi_v^{km}(r)$ are a finite set of OPWs and the Hamiltonian H is given by

$$H = -\frac{\hbar^2}{2m}\nabla^2 + V(r) \tag{3.44}$$

where $V(r)$ is the full crystal potential.

It is not immediately obvious why the solutions of eq. (3.43) should be NFE-like, requiring only a small number of OPW states to determine the valence energy levels. When we expand the Schrödinger equation, by substituting eq. (3.40) into (3.43), we find two different types of large terms in the expansion, the first type associated with the full potential acting on the plane wave part of the OPWs, $V(r)e^{ik \cdot r}$, and the second type associated with the full Hamiltonian acting on the core state part of the OPWs, $H \sum a_{ik}\phi_{ci}(r) = \sum \varepsilon_{ci}a_{ik}\phi_{ci}(r)$.

Fortunately, the effect of the two types of terms largely cancel each other. If we consider a nucleus centred at the origin ($r = 0$), then the first term, $V(r)e^{ik \cdot r}$, is large and negative near $r = 0$, while the second term, $\sum \varepsilon_{ci}a_{ik}\phi_{ci}(r)$ can be shown to be large and positive. We can re-write eq. (3.43) to take advantage of this cancellation, effectively replacing the true potential $V(r)$ by a much weaker pseudopotential $V_{ps}(r)$ and the true basis states $\phi_v^{km}(r)$ by plane-wave pseudobasis states, $e^{ik(m)x}$. This gives a new, so-called pseudoHamiltonian, whose eigenvalues are identical to the valence eigenvalues of the true Hamiltonian, and with the pseudo-Hamiltonian basis states given by linear combinations of a small number of plane wave pseudofunctions.

To illustrate how the two types of term cancel, leaving a weaker pseudopotential with a smooth pseudowavefunction, we take an example based on the K–P model (Weaire and Kermode 1985). We consider a series of square wells and barriers of equal width ($a = b = 6$ Å), with the zero of energy at the bottom of the well and the barrier of height $V_0 = 2.0$ eV (fig. 3.16a). The solid lines in fig. 3.16(b)–(d) show the exact wavefunctions for the three lowest zone centre ($q = 0$) states, which have energies $E_1 = 0.47$ eV, $E_2 = 1.82$ eV and $E_3 = 2.50$ eV.

The dashed line in fig. 3.16(d) shows that the exact wavefunction for the third state can be well approximated by a single plane wave with $k = 0$ (fig. 3.16e) orthogonalised to the lowest energy level:

$$\phi_3^{OPW}(x) = \frac{1}{\sqrt{2a}} - \psi_1(x) \int_{-a}^{a} dx' \frac{1}{\sqrt{2a}} \psi_1(x')$$

$$= 1/(2a)^{\frac{1}{2}} - \alpha_1 \psi_1(x) \qquad (3.45)$$

We see from fig. 3.16(d) that $\phi_3^{OPW}(x)$ is then a good variational estimate of the true wavefunction. We can, therefore, substitute the orthogonalised state (eq. (3.45)) into the exact Schrödinger equation, $H\psi = E\psi$, to estimate the energy, E_3, of the third state. We show below how $H\phi_3^{OPW} = E_3\phi_3^{OPW}$ can be re-arranged to give a new equation with the same eigenvalue, E_3, but with a weaker pseudopotential and smoother wavefunction. To do so,

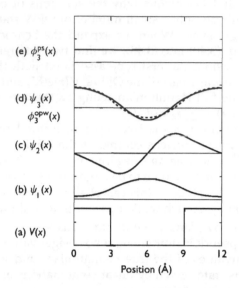

(e) $\phi_3^{ps}(x)$

(d) $\psi_3(x)$
    $\phi_3^{OPW}(x)$

(c) $\psi_2(x)$

(b) $\psi_1(x)$

(a) $V(x)$

0   3   6   9   12
Position (Å)

Figure 3.16 (a) The K–P potential with $a = b = 6\,\text{Å}$, and barrier height $V_0 = 2.0\,\text{eV}$. The solid lines in (b)–(d) show the exact wavefunctions for the three lowest zone centre states in this potential. The exact wavefunction $\psi_3(x)$ for the third state (solid line in (d)) is well approximated by the dashed line $\phi_3^{OPW}(x)$ in (d), formed from a single plane wave (with wavevector $k = 0$, (e)) orthogonalised to the lowest state, as described in the text.

we first expand the left-hand side, $H\phi_3^{OPW}(x)$, as the sum of several terms:

$$H\phi_3^{OPW}(x) = H\frac{1}{\sqrt{2a}} - H\alpha_1\psi_1(x)$$

$$= -\frac{\hbar^2}{2m}\frac{d^2}{dx^2}\frac{1}{\sqrt{2a}} + V(x)\frac{1}{\sqrt{2a}} - \alpha_1 E_1\psi_1(x) \qquad (3.46)$$

while the right-hand side is given by

$$E\phi_3^{OPW}(x) = E\left(\frac{1}{\sqrt{2a}} - \alpha_1\psi_1(x)\right) \qquad (3.47)$$

We can equate eqs (3.46) and (3.47), the two sides of the true Schrödinger equation, and then re-arrange terms to give a modified second order wave equation, where we set $\phi_3^{PS}(x) = 1/\sqrt{2a}$ to give

$$\left[-\frac{\hbar^2}{2m}\frac{d^2}{dx^2} + V(x) + \alpha_1(E_3 - E_1)\sqrt{2a}\psi_1(x)\right]\phi_3^{PS}(x) = E_3\phi_3^{PS}(x)$$

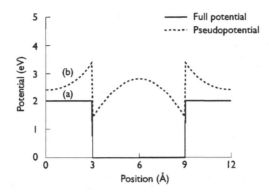

Figure 3.17 Comparison of (a) the original K–P potential and (b) of the (weaker) pseudopotential introduced through eq. (3.49) to determine the energy of the third state in the K–P potential of fig. 3.16.

that is,

$$H_{ps}\phi_3^{ps}(x) = E_3\phi_3^{ps}(x) \qquad (3.48)$$

where the pseudopotential, $V_{ps}(x)$ is made up of two large terms of opposite sign, which approximately cancel,

$$V_{ps}(x) = V(x) + \alpha_1(E_3 - E_1)\sqrt{2a}\psi_1(x) \qquad (3.49)$$

as illustrated in fig. 3.17. We can then use the variational method, with $\phi_3^{ps}(x) = (2a)^{-1/2}$ as trial function, to estimate the energy of the third zone centre state as 2.52 eV, the average value of the 'smooth' potential in fig. 3.17. Figure 3.18 illustrates the general applicability of the pseudopotential method, showing that the trial function, $\phi_3^{ps}(x)$, provides a very good estimate of E_3 over a wide range of barrier heights.

It must be admitted that the potential in fig. 3.17 is not significantly smoother than that of the original K–P potential. This however is because the original K–P potential was reasonably smooth anyway. Figure 3.16(e) shows however that we have achieved the smoothest possible pseudo-wavefunction, $\phi_3^{PS}(x)$. When the pseudopotential method is applied to atoms in solids, it does genuinely smooth the potential: we can replace the $1/r$ singularity at each nucleus by a slowly varying, or even constant potential.

Hundreds if not thousands of man-years of research have already gone into the development and refinement of the pseudopotential method, so that the most sophisticated, *ab initio*, pseudopotentials can now predict the static, ground-state, properties of very complex structures, and are also used to investigate dynamic behaviour such as atomic diffusion in solids, crack propagation at grain boundaries and phase transitions.

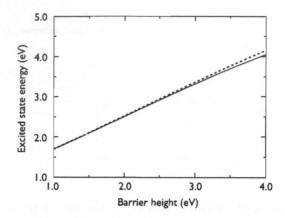

Figure 3.18 Energy of the third state in the K–P potential where $a = b = 6\text{Å}$, calculated as a function of the barrier height V_0. Solid line: exact energy; dashed line: energy estimated using the pseudopotential method.

We have only scratched the surface here in describing some of the different general methods used to determine the electronic properties of solids. We develop some of these ideas further in the next chapter, and then use them to investigate some of the key features of semiconductors, magnetism and superconductivity in the remainder of the book.

References

Ashcroft, N. W. and N. D. Mermin (1976) *Solid State Physics*, Holt, Rinehart and Winston, New York.
Harrison, W. A. (1989) *Electronic Structure and the Properties of Solids: The Physics of the Chemical Bond*, Dover Publications, New York.
Ibach, H. and H. Lüth (1995) *Solid State Physics*, Springer-Verlag, New York.
Pettifor, D. M. (1995) *Bonding and Structure of Molecules and Solids*, Oxford University Press.
Weaire, D. and J. P. Kermode (1985) *phys. stat. sol. (b)* **127**, K143.

Problems

3.1   Consider a linear chain of atoms, distance L apart, for which the band structure is given by

$$E_{sq} = E_s + 2V \cos(qL) \tag{3.50}$$

where E_s is the self-energy of the single orbital on an isolated atom, and V is the nearest neighbour interaction in the chain. Suppose that each orbital can accommodate two electrons. Calculate from eq. (3.50) the average band-structure energy gained per atom if there are y electrons per atom on the linear chain.

3.2   The linear chain of atoms in problem 3.1 undergoes a distortion so
that the even-numbered atoms move to the right, doubling the unit
cell size to $2L$, and leading to an interaction of strength $V + \Delta V$
between the $2n$th and $(2n + 1)$th site, and an interaction of strength
$V - \Delta V$ between the $2n$th and $(2n - 1)$th site. By writing the wavefunction within a given unit cell as a linear combination of the two
orbitals in that cell, and then applying Bloch's theorem, show that
the band structure in the distorted lattice is given by

$$E_{sq} = E_s \pm 2\sqrt{V^2 \cos^2(qL) + (\Delta V)^2 \sin^2(qL)} \qquad (3.51)$$

and hence show that the distortion can be regarded as opening an
energy gap of magnitude $4(\Delta V)$ at $q = \pi/2L$ in the band structure of
problem 3.1. Estimate the average band-structure energy gained per
atom when this distortion, referred to as a Peierls distortion, occurs
in a linear chain with one electron per atom on the chain.

3.3   Show for the K–P potential of eq. (3.31) that the magnitude of the
nth energy gap at $q = \pm\pi/L$ is given in the NFE model by

$$E_g = \frac{2V_0}{(2n-1)\pi} \sin\left(\frac{(2n-1)\pi b}{L}\right) \qquad (3.39b)$$

while the nth gap at $q = 0$ is given by

$$E_g - \frac{V_0}{n\pi} \sin(2n\pi b/L) \qquad (3.39c)$$

3.4   Show for the K–P potential of eq. (3.31) that the NFE wavefunction,
$\psi_{n,l}(x)$, of the lower state at the nth gap at $q = \pm\pi/L$ is given by

$$\psi_{n,l} = \sqrt{\frac{2}{L}} \sin\left(\frac{(2n+1)\pi x}{L}\right)$$

while that of the upper state, $\psi_{n,u}(x)$, is given by

$$\psi_{n,u} = \sqrt{\frac{2}{L}} \cos\left(\frac{(2n+1)\pi x}{L}\right)$$

Hence justify why the NFE method provides a better estimate of the
energy, to larger values of b, for the lower state than for the upper
state.

3.5   We can improve the accuracy of the NFE method by including more
basis states in the NFE calculation of the K–P band structure. We
see from the previous question that the four lowest basis states at

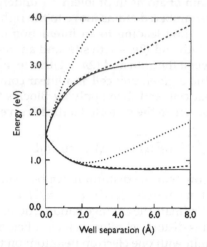

Figure 3.19 Calculated variation of the two lowest band edge energies at $q = \pi/L$ for the K–P potential with well width $a = 5\,Å$ and barrier height $V_0 = 5\,eV$. The solid lines show the exact results. The dotted lines show the energies calculated using the simplest form of the NFE method (eq. (3.38)). The dashed lines include extra basis states in the NFE method, as described in problem 3.5.

$q = \pm\pi/L$ can be chosen as $\alpha \sin(\pi x/L)$, $\alpha \cos(\pi x/L)$, $\alpha \sin(3\pi x/L)$ and $\alpha \cos(3\pi x/L)$, where $\alpha = (2/L)^{1/2}$. By considering separately the basis functions which are even and odd about the origin, show that when we include the extra basis states, the two lowest energy states at $q = \pi/L$ are given by

$$E = \frac{5h^2}{8mL^2} + \frac{bV_0}{L} \mp \frac{V_0}{2\pi}\left(\sin\frac{\pi b}{L} + \frac{1}{3}\sin\frac{3\pi b}{L}\right)$$

$$-\sqrt{\left[\frac{h^2}{mL^2} \mp \frac{V_0}{2\pi}\left(\frac{1}{3}\sin\frac{3\pi b}{L} - \sin\frac{\pi b}{L}\right)\right]^2 + \frac{V_0^2}{\pi^2}\left(\sin\frac{\pi b}{L} \mp \frac{1}{2}\sin\frac{2\pi b}{L}\right)^2}$$

Figure 3.19 shows how the inclusion of the extra basis states allows the NFE method to work to considerably larger values of b.

3.6   Consider the complex wave of amplitude

$$\psi(x,y,t) = \exp(i(k_x x + k_y y - \omega t))$$

where $k_x = k\cos\theta$ and $k_y = k\sin\theta$. Calculate the wavelength λ, the phase velocity v and the direction of motion of this wave. Consider the square region

$$0 < x < L$$
$$0 < y < L$$

and let $k_x = k_y = 8\pi/L$. Draw out (i) for $t = 0$ and (ii) for $t = \pi/\omega$, the lines along which $\psi(x, y, t) = 1$. Calculate the repeat distance of the wave along the x-direction, the y-direction, and its direction of motion.

3.7 Show that $a_1 = (0, a/2, a/2)$, $a_2 = (a/2, 0, a/2)$ and $a_3 = (a/2, a/2, 0)$ are a set of primitive (i.e. basis) vectors for the FCC lattice, meaning any lattice site R can be generated as a linear combination of an integer times each of these three vectors:

$$R = n_1 a_1 + n_2 a_2 + n_3 a_3$$

3.8 Each reciprocal lattice vector G for a given crystal lattice defines the wavevector of a plane wave which has the same periodicity as the lattice, that is if R is a vector joining the same point in two different unit cells of the crystal lattice, then we require $e^{iG \cdot R} = 1$, or $G \cdot R = 2\pi n$, where n is an integer. A particular crystal lattice is defined by the three primitive vectors, a_1, a_2 and a_3. Show that the reciprocal lattice can then be defined in terms of the three primitive vectors b_1, b_2 and b_3, where

$$b_1 = \frac{2\pi(a_2 \times a_3)}{a_1 \cdot (a_2 \times a_3)}; \quad b_2 = \frac{2\pi(a_3 \times a_1)}{a_1 \cdot (a_2 \times a_3)}; \quad b_3 = \frac{2\pi(a_1 \times a_2)}{a_1 \cdot (a_2 \times a_3)}$$

3.9 Using the FCC basis vectors of problem 3.7, calculate the basis vectors for the FCC reciprocal lattice, and hence show that the reciprocal lattice of an FCC lattice is a BCC lattice (and vice versa).

3.10 A 2-D triangular lattice is defined by the basis vectors $a_1 = (a, 0)$; $a_2 = (a/2, \sqrt{3}a/2)$. Determine the reciprocal lattice basis vectors for this triangular lattice, and hence show that the reciprocal lattice of a triangular lattice is itself a triangular lattice.

3.11 The triangular lattice of problem 3.10 contains one atom per unit cell, which has a single orbital with self-energy E_s. Show that if this orbital has an interaction $V(V < 0)$ with each of its six neighbours, then the triangular lattice band structure is given by

$$E_{sk} = E_s + 2V \left(\cos(k_x a) + 2 \cos(k_x a/2) \cos(\sqrt{3}k_y a/2) \right)$$

Show that the lower and upper band edges are then at energies $E_{sL} = E_s - 6|V|$ and $E_{sU} = E_s + 2|V|$, respectively. Justify why the most strongly bonding state, E_{sL}, is in this case shifted considerably further down in energy than E_{sU} is shifted upwards.

Chapter 4

Band structure and defects in bulk semiconductors

4.1 Introduction

The exact features of the band structure of semiconductor materials play a crucial role in determining their electronic and optoelectronic properties and their usefulness for devices such as transistors, microwave oscillators, lasers, light emitting diodes, and detectors. For this reason a great deal of effort has been given to finding ways to engineer the band structure, including the use of alloys, heterojunctions, quantum confinement, and strain. Some of the most advanced devices now employ all of these techniques. In this chapter, we examine some of the main features of the electronic band structure of bulk semiconductors, including the influence of defect atoms on the band structure. The next chapter discusses the physics and applications of some of the more advanced low-dimensional semiconductor structures.

Figure 4.1 shows the band structures of bulk gallium arsenide (GaAs) and silicon (Si), which are two of the most important semiconductors. We are often particularly interested in the bands near the energy gap, E_g, which separates the filled valence (bonding) bands from the empty conduction (anti-bonding) bands. As can be seen in fig. 4.1(a), the lowest energy state in the GaAs conduction band, E_c, is at the Γ point, at the Brillouin zone centre, directly above the highest energy state in the valence band, E_v, and so GaAs is called a *direct* gap semiconductor. This has the very important consequence that an electron which has relaxed to its lowest possible energy in the conduction band can recombine directly with a hole at the top of the valence band simply with the emission of a photon. (Note that on the scale of fig. 4.1 a photon has negligible momentum.) Semiconductors with a direct band gap, therefore, make efficient light emitting devices such as lasers and light emitting diodes.

Figure 4.1(b) shows the band structure of Si. It is described as an *indirect* band gap semiconductor, because the bottom of the conduction band is not directly above the valence band maximum at Γ. In fact there are six equivalent conduction band minima in Si, along the six equivalent Δ directions,

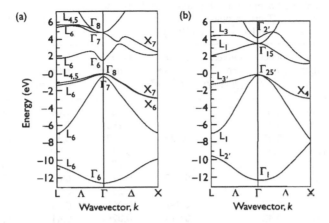

Figure 4.1 The band structure of (a) GaAs and (b) Si, as calculated by Chelikowsky and Cohen (1976) using the empirical pseudopotential method (© 1976 by the American Physical Society). The zero of energy is at the valence band maximum, E_v. The conduction band minimum is at energy $E_c \approx 1.5\,eV$ in GaAs and at $E_c \approx 1.1\,eV$ in Si, giving an energy gap of magnitude $E_g \approx 1.5$ and 1.1 eV respectively.

near to the X points. It is possible for an electron at the bottom of a Δ minimum to recombine with a hole at the top of the valence band with the emission of a photon but, in order to conserve momentum, the process must be accompanied by the emission or absorption of a phonon. (By analogy with photon and electron, a phonon is the name given to a quantised lattice vibration. Because the mass of a nucleus is very large compared to that of an electron, phonons can have negligible energy but significant momentum on the scale of fig. 4.1.) Because light emission from indirect gap materials requires the simultaneous occurrence of two processes, it is intrinsically less efficient than in the direct gap case. Besides Si, the other main group IV semiconductors are also indirect gap materials, with the conduction band minimum near the X point in diamond (C) and at the L point in germanium. Many III–V semiconductors are direct gap and therefore suitable for optoelectronic applications. Table 4.1 lists some of the important material properties of several technologically interesting semiconductor materials.

The valence band maximum is always at the Γ point in tetrahedrally bonded semiconductors. If we ignored the interaction between each electron's spin and its orbital motion (the spin–orbit interaction), the valence band maximum would consist of three degenerate p-like bonding states (six states in fact when we allow for electrons with spin 'up' and spin 'down'). However, consider the electrons in the outermost p states of an

Table 4.1 Selected properties of group IV and III–V tetrahedrally bonded semiconductors. [Data compiled from Madelung (1982); estimates of GaN effective masses from Meney *et al.* (1996)]

	Bond length $d(Å)$	Direct gap $E_g(eV)$	Indirect gap $E_g^{ind}(eV)$	Electron relative mass m_{cr}^*	LH relative mass m_{lr}^*	HH relative mass m_{hr}^*	Dielectric constant ε_r
C	1.54	6.5	5.5		0.36	1.08	5.7
Si	2.35	4.2	1.17		0.153	0.54	12.1
Ge	2.44	0.90	0.74	0.038	0.043	0.35	16.5
GaN	1.94	3.5		0.15	0.21	0.90	10
GaP	2.36	2.88	2.35		0.17	0.67	9.1
InP	2.54	1.42		0.079	0.12	0.6	9.6
AlAs	2.43	3.13	2.23		0.26	0.5	9.1
GaAs	2.45	1.52		0.067	0.082	0.45	12.8
InAs	2.61	0.42		0.023	0.026	0.41	12.3
AlSb	2.66	2.32	1.69		0.11	0.5	10.2
GaSb	2.65	0.81		0.041	0.05	0.4	14.4
InSb	2.8	0.235		0.0139	0.018	0.4	15.7

Table 4.2 Spin–orbit splitting energy (in meV) at the valence band maximum in III–V zinc-blende (upper part of table) and group IV tetrahedrally bonded semiconductors. (III–V data from Krijn (1991); nitride data from Wei and Zunger (1996) and group IV data from Madelung (1982))

	N	P	As	Sb
Al	19	70	280	650
Ga	15	80	340	820
In	6	110	380	810

C	Si	Ge
6	45	290

isolated atom. The atomic p states there are also degenerate when we ignore the electron spin. In reality, the electrons experience an additional potential due to the interaction of the spin magnetic moment with the vector product of the velocity and electric field. This effect, spin–orbit coupling, splits the p states in an isolated atom. It persists in solids and splits the degeneracy of the valence band maximum (as well as splitting other degenerate states in the band structure). Two of the six p states at the valence band maximum are split off from the other four, and shifted down to lie at an energy E_{so} below the valence band maximum. Table 4.2 lists the spin–orbit splitting, E_{so}, for the group IV and III–V semiconductors. It can be seen how the magnitude of the spin–orbit splitting, E_{so}, increases with atomic number and is determined primarily by the group V element in III–V materials.

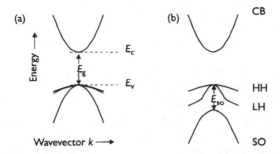

Figure 4.2 The band structure of a direct gap bulk semiconductor in the vicinity of the band gap and near the centre of the Brillouin zone. (a) Assuming the spin–orbit splitting energy, $E_{so} = 0$, and (b) for finite E_{so}. The lowest conduction state (at energy E_c) is separated by the band-gap energy E_g from the highest valence state, at energy E_v. The labels CB, HH, LH and SO indicate the conduction band, heavy-hole, light-hole and spin-split-off bands, respectively.

Figure 4.2 illustrates the band structure of a direct gap bulk semiconductor in the vicinity of the energy gap. Because the energy varies quadratically with wavevector k near the zone centre, we can, for instance, describe the conduction band dispersion near the band edge, $E_{CB}(k)$, by

$$E_{CB}(k) = E_c + \frac{\hbar^2 k^2}{2m_c^*} \qquad (4.1a)$$

where we introduce the electron effective mass, m_c^* as the constant of proportionality linking E to k. The concept of a carrier 'effective mass' turns out to be a very useful idea. Although it is introduced phenomenologically here, we shall show in Section 4.3 that for many purposes we can treat, for example, an electron at the bottom of the conduction band as if it were a particle in free space, with a mass m_c^*.

The two highest valence bands are referred to as the heavy-hole (HH) and light-hole (LH) bands, because the dispersion of the highest band always varies slowly away from the zone centre, and can be described as if the highest band has a heavy effective mass, m_{HH}^*

$$E_{HH}(k) = E_v - \frac{\hbar^2 k^2}{2m_{HH}^*} \qquad (4.1b)$$

while the dispersion of the next band varies considerably more rapidly near the zone centre, as if the band had a light effective mass, m_{LH}^*

$$E_{LH}(k) = E_v - \frac{\hbar^2 k^2}{2m_{LH}^*} \qquad (4.1c)$$

The third valence band is generally called the spin-split-off band. The lowest conduction band in a direct gap semiconductor always has a low effective mass; m_c^* is always smaller than m_{LH}^*, as can be seen in Table 4.1. It can also be observed that m_c^* and m_{LH}^* increase with increasing energy gap, E_g.

Clearly, these trends in semiconductor band structure must have a physical basis, which could be elucidated using either the tight-binding or pseudopotential method introduced in Chapter 3. In practice, however, the observed trends are best explained using a semi-empirical technique, called $k \cdot p$ theory, which will also prove very useful when considering topics such as doping and low-dimensional semiconductor structures later in this and in the next chapter.

4.2 $k \cdot p$ theory for semiconductors

$k \cdot p$ theory is a perturbation method, whereby if we know the exact energy levels at one point in the Brillouin zone (say $k = 0$, the Γ point) then we can use perturbation theory to calculate the band structure near that k value. The zone centre energy levels have been determined experimentally for many semiconductors, as listed in Tables 4.1 and 4.2. A general introduction to first and second order perturbation theory is given in Appendix C, while the $k \cdot p$ model is summarised here and described in more detail in Appendix E.

From Bloch's theorem, Schrödinger's equation in a periodic solid can be written as

$$H_0 \psi_{nk}(r) = E_{nk} \psi_{nk}(r) \tag{4.2}$$

where H_0 is the k-independent Hamiltonian acting on ψ_{nk}, the k-dependent wavefunction $e^{ik \cdot r} u_{nk}(r)$ associated with the state with energy E_{nk}. We show in Appendix E how it is possible to transform eq. (4.2) to give a k-dependent Hamiltonian, $H_k = e^{-ik \cdot r} H_0 e^{ik \cdot r}$, acting on the wavefunction $u_{nk}(r)$. H_k is given by

$$\begin{aligned} H_k &= H_0 + H_k' \\ &= H_0 + \frac{\hbar}{m} k \cdot p + \frac{\hbar^2 k^2}{2m} \end{aligned} \tag{4.3}$$

where p is the momentum operator, $p = -i\hbar \nabla$. We expect for small values of k that the energy levels E_{nk} of H_k will be very close to the known energy levels E_{n0} of H_0. We can then use standard perturbation theory to estimate that the dispersion along the i-direction near the band edge at energy E_{n0}

is given by

$$E_{nk} = E_{n0} + \frac{\hbar^2 k^2}{2m} + \frac{\hbar^2 k^2}{m^2} \sum_{n' \neq n} \frac{|i \cdot p_{nn'}|^2}{E_{n0} - E_{n'0}} \qquad (4.4a)$$

where i is a unit vector pointing along the direction i, and $p_{nn'}$ is the momentum matrix element linking the nth and n'th basis state of H_0 (see Appendix E). From the definition of m^* above, we can then estimate the effective mass along the direction i in the nth band, m_i^*, as

$$\frac{1}{m_i^*} = \frac{1}{m} + \frac{2}{m^2} \sum_{n' \neq n} \frac{|i \cdot p_{nn'}|^2}{E_{n0} - E_{n'0}} \qquad (4.4b)$$

For real semiconductors, we should take account of the spin–orbit interaction and also of band degeneracies, neither of which were included in the derivation of eq. (4.4). The practical application of the $k \cdot p$ model, therefore, looks rather complicated. However, two main reasons help to ensure its usefulness (Kane 1966; Bastard 1988):

1   In most cases, we only have to deal with a very small number of bands, which are close to each other in energy, and can ignore higher and lower bands, where the denominator in the summation terms of eq. (4.4) is large.

2   When we restrict the number of bands in eq. (4.4), and then start to fit experimental data, such as measured effective masses, m^*, and measured energy gaps $E_{n0} - E_{n'0}$ it is found that the momentum matrix elements $p_{nn'}$ are remarkably constant between different semiconductors, with $p_{nn'}$ identically zero for many pairs of bands n and n'.

We can consider a very simple (but not totally unrealistic) model of a direct-gap semiconductor, with zero spin–orbit splitting. The conduction band consists of a single s-like anti-bonding state at $k = 0$, while the top of the valence band has pure p-like symmetry, so is threefold degenerate. We choose the three p states to point along the three crystal axes, and label them p_x, p_y and p_z, respectively.

It can be shown from symmetry considerations that the matrix elements p_{ij} linking any pair of the p states are identically zero, while the momentum matrix element linking s with each p level points along the direction of that p state, so that $p_{sx} = Pi$ say, with $p_{sy} = Pj$ and $p_{sz} = Pk$, where i, j, and k are unit vectors along the three crystal axes.

If we assume that the wavevector k points along the x-direction, then the matrix elements $k \cdot p_{nn'}$ corresponding to $k \cdot p_{sy}$ and $k \cdot p_{sz}$ are identically

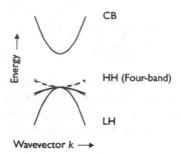

Figure 4.3 Band structure of a direct gap tetrahedral semiconductor calculated using the 4-band **k · p** model (1 conduction; 3 valence bands) and assuming $E_{so} = 0$. The dashed lines show the HH dispersion in the 4-band model; the solid lines immediately below indicate how interactions with higher conduction bands push the HH bands downwards.

zero. Therefore, eq. (4.4) gives for the p_y and p_z bands that

$$\frac{1}{m_y^*} = \frac{1}{m_z^*} = \frac{1}{m} \tag{4.5}$$

along the x-direction. These are the two HH bands, whose effective mass m_{HH}^* then equals the free-electron mass, m, as illustrated by the dashed lines in fig. 4.3. Interactions with higher conduction bands, which we have ignored, will tend to push the bands further downwards (solid lines), giving them the typical dispersion observed for HH bands in figs 4.1 and 4.2. The LH band arises from the interaction between the s and p_x states, for which $|k \cdot p_{sx}| = kP$, so that eq. (4.4b) then gives the LH effective mass, m_{LH}^*, as

$$\frac{1}{m_{LH}^*} = \frac{1}{m_x^*} = \frac{1}{m} - \frac{2P^2}{m^2 E_g} \tag{4.6}$$

along the x-direction. For the conduction band, the sign of the energies is reversed in the denominator, and we have

$$\frac{1}{m_c^*} = \frac{1}{m} + \frac{2P^2}{m^2 E_g} \tag{4.7}$$

In fact, the value of the second term on the right-hand side of eqs (4.6) and (4.7) is much larger than the first term, $1/m$, so that $m_c^* \approx |m_{LH}^*|$ for zero spin–orbit interaction. We can rewrite eq. (4.7) as

$$\frac{m}{m_c^*} = 1 + \frac{E_p}{E_g} \tag{4.8}$$

where we have set $E_p = 2P^2/m$. When we include the spin–orbit interaction, the expressions for the valence and conduction effective masses are modified to account for the revised interactions between the conduction band and the spin-split-off band, with the effective masses for the conduction and three valence bands then given by

$$\frac{m}{m_c^*} = 1 + \frac{E_p}{3}\left(\frac{2}{E_g} + \frac{1}{E_g + E_{so}}\right) \tag{4.9a}$$

$$\frac{m}{m_{LH}^*} = 1 - \frac{2E_p}{3E_g} \tag{4.9b}$$

$$\frac{m}{m_{so}^*} = 1 - \frac{E_p}{3(E_g + E_{so})} \tag{4.9c}$$

and $m/m_{HH}^* = 1$, as before.

We can use eq. (4.9a) along with the quoted values of E_g, E_{so}, and m_c^* in Tables 4.1 and 4.2 to estimate the value of E_p for a range of materials. This is left as an exercise to the end of the chapter, where it can be observed that the magnitude of E_p is reasonably constant throughout the III–V semiconductors. It is further possible, using eq. (4.9b) and the calculated values of E_p to estimate the magnitude of the LH masses, m_{LH}^*, which are then observed to be in generally good agreement with the experimentally determined values listed in Table 4.1.

The above calculations serve to show the value of $k \cdot p$ theory when applied to tetrahedrally bonded semiconductors. As a further example, we consider the highest valence and lowest conduction band along the Δ and Λ directions, linking X and L respectively to Γ. There is no $k \cdot p$ interaction between the conduction and valence band along these directions. Therefore, the highest valence and lowest conduction band are approximately parallel for much of Γ–X and Γ–L, as illustrated in fig. 4.1. The effective mass along this (longitudinal) direction, m_l^*, is then large at the X and L points (of the order of the free electron mass), and consequently there is a much larger density of states associated with the lowest X and L conduction band minima compared to the lowest Γ minimum.

By contrast, there is a large $k \cdot p$ interaction in the transverse direction at X and L (moving perpendicular to the Δ and Λ directions), so that the transverse effective mass, m_t^* is then comparatively small at X and L. Surfaces of constant energy near X and L are then described by ellipsoids, as illustrated in fig. 4.4.

4.3   Electron and hole effective masses

We introduced the electron effective mass phenomenologically in Section 4.1, claiming, for example, that for many purposes we can treat

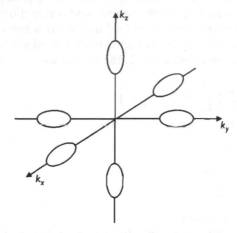

Figure 4.4 Surfaces of constant energy are described by ellipsoids near the lowest conduction band X points. This occurs because there is a large effective mass m_l^* along the Δ directions (pointing towards Γ, the Brillouin zone centre), with a smaller effective mass, m_t^* perpendicular to the Δ direction.

an electron at the bottom of the conduction band as if it were a particle in free space, with an effective mass m_c^*. If we apply an external electric field E, such that the electron experiences a force $F = -eE$ this implies the electron will move with an acceleration a given by $m_c^* a = -eE$; that is,

$$a = \frac{F}{m_c^*} = \frac{-eE}{m_c^*} \tag{4.10}$$

For small effective mass, the electron then accelerates more rapidly in the solid than in free space. This is at first surprising, but reflects the fact that the electron is acted on not only by the external field E but also by the periodic field of the lattice. If we took explicit account of both fields in discussing the dynamics of the electron it would exhibit its ordinary mass.

In order to derive the correct form for the electron effective mass, we consider an electron represented by a wave packet near the bottom of the conduction band, so that the electron velocity is then given by the group or energy velocity, v_g, defined as

$$v_g = \frac{d\omega}{dk} = \frac{1}{\hbar} \frac{dE}{dk} \tag{4.11}$$

where k is the Bloch wavevector associated with the electron state of energy E, and for simplicity we only consider motion in one dimension. In the applied field E, the electron experiences a force F such that its energy E

increases by

$$\delta E = F\,\delta x = Fv_g\,\delta t$$

$$= F\frac{1}{\hbar}\frac{dE}{dk}\delta t \tag{4.12}$$

But we can also relate δE to the change in wavevector δk, as

$$\delta E = \frac{dE}{dk}\,\delta k \tag{4.13}$$

By comparing the right hand sides of eqs (4.12) and (4.13) we then find that

$$F = \hbar\frac{dk}{dt} \tag{4.14}$$

This holds irrespectively of whether the electron is in free space or in a periodic potential. We can use eq. (4.11) to determine the electron acceleration a as

$$a = \frac{dv_g}{dt} = \frac{1}{\hbar}\frac{d^2E}{dk\,dt} = \frac{1}{\hbar}\frac{d^2E}{dk^2}\frac{dk}{dt} \tag{4.15}$$

Substituting eq. (4.14) into eq. (4.15) we find

$$a = \frac{1}{\hbar^2}\frac{d^2E}{dk^2}F \tag{4.16}$$

By comparison with Newton's law (eq. (4.10)), we see that the electron then behaves as if it has an effective mass, m^*_{eff} given by

$$\frac{1}{m^*_{eff}} = \frac{1}{\hbar^2}\frac{d^2E}{dk^2} \tag{4.17}$$

This broadens the definition of effective mass for a parabolic band in eq. (4.1) to the more general case of a non-parabolic band dispersion.

We see from eq. (4.17) and elsewhere that an electron can accelerate more rapidly in a solid than in free space. How does this happen? We note that in free space the potential energy is constant and so only the kinetic energy varies with wavevector k. By contrast, in a solid the total energy of any state is the sum of a k-dependent kinetic and a k-dependent potential energy contribution. We use the NFE method to show in the problems at the end of this chapter that the average potential energy seen by a conduction electron can decrease as the electron wavevector k shifts away from the conduction band minimum. This decrease in average potential energy with increasing k then enables the electron to accelerate more rapidly in the periodic solid than in free space.

We also see from eqs (4.17) and (4.9b) that an electron near the top of the valence band will behave as if it has a *negative* effective mass, implying that the electron will be accelerated in the direction *opposite* to an applied force. This is both surprising, and counter-intuitive.

Why can an electron be accelerated against the applied force? The answer is again because the electron is acted on both by the external field and the periodic field of the lattice. From eq. (4.14) we have, both in free space and in a periodic solid, that an applied force acting in the positive direction will lead to the wavevector k associated with a given state becoming more positive. Because both kinetic *and* potential energy depend on k in the periodic solid, there can be several ranges of k where the kinetic energy decreases as the magnitude of k increases. Relative to the externally applied force, the electron then behaves as if it has a negative effective mass. In addition, if we consider a one-dimensional structure with period a then we have seen from Bloch's theorem that we can associate the wavenumber k or $k - 2\pi/a$ with each state, so that the wavevectors π/a and $-\pi/a$ are equivalent. The periodic potential then enables the Bragg reflection of the wave-like electrons in the direction opposite to the applied force.

It can be shown (see problem 4.10) that the electrons in a filled band experience zero net acceleration: effectively for every electron accelerated to the left, another is accelerated to the right. Hence a filled band will not contribute to the current flow in a solid.

Suppose that we now excite one electron from near the top of a filled valence band to near the bottom of an empty conduction band. The single excited conduction band electron will then behave, as already discussed, like a free carrier with positive effective mass given by eq. (4.17).

How will the almost filled valence band respond to an external force, F? Before it was excited into the conduction band, the electron near the valence band maximum had a negative effective mass, implying that in its contribution to the total current it was being accelerated against the applied force. (Because the electron charge is negative, the electron was, therefore, being accelerated along the direction of current flow, with its negative charge, therefore, tending to reduce the net current.) Because the total current is zero in a filled band, when the negatively charged electron is removed, the net current due to the remaining electrons must be equal and opposite to that associated with the missing electron. The remaining electrons then experience a net acceleration in the direction of the applied force, that is, against the direction of current flow. Because the negatively charged electrons are being accelerated against the direction of current flow, their motion, therefore, tends to increase the net current.

When a negatively charged electron is removed from an otherwise filled band, the empty state left behind is often referred to as a 'hole' state. We argued above that removing an electron with negative effective mass increases the net current flow in the direction of the applied field E. We can, therefore, treat this contribution to the current as though it were due to a positively charged carrier, referred to as a hole, with a (positive) effective mass, m_h^* equal and opposite to that of the missing electron:

$$\frac{1}{m_h^*} = -\frac{1}{\hbar^2}\frac{d^2E}{dk^2} \tag{4.18}$$

We then have the choice to describe the incomplete valence band either in terms of the (very many) filled electron states or the (few) hole states. The latter description is particularly appropriate when considering semiconductor valence bands and will be used later in this and in the next chapter.

4.4 Trends in semiconductors

4.4.1 Alloying

We saw in fig. 2.16 how the energy gaps, E_g, and bond lengths, d, of the III–V semiconductors span a wide range, from $d = 2.81$ Å and $E_g = 0.17$ eV for InSb to $d = 1.89$ Å and $E_g = 6.2$ eV for AlN. It is, in addition, possible to form a wide range of semiconductor alloys, such as $Al_xGa_{1-x}As$, where we have replaced a fraction, x, of the group III gallium atoms by aluminium atoms. When we do so, the lattice constant and band structure of tetrahedrally bonded semiconductors generally vary smoothly as the alloy composition is varied. This is in fact also the case for the electronic structure of most solid alloys. Figure 4.5 illustrates the variation

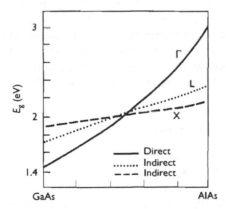

Figure 4.5 Variation of the energy gaps at 300 K between the valence band maximum and conduction band minima as a function of aluminium composition, x, in $Al_xGa_{1-x}As$. E_Γ is the direct energy gap, while E_L and E_X are the energy gaps to the lowest conduction band states near the L and X points, respectively. (After Casey and Pannish, 1978.)

of the different energy gaps in $Al_xGa_{1-x}As$, as a function of aluminium composition, x. It can be seen that $Al_xGa_{1-x}As$ is a direct gap semiconductor for low aluminium concentrations, with the energy gap becoming indirect for $x > 0.45$, beyond which the lowest conduction states lie along the Δ direction, near the X point. This smooth variation of the bond length and electronic properties enables the band structure engineering described in the next chapter, whereby it is possible to tailor the band gap and related properties for a wide range of applications.

4.4.2 Hydrostatic pressure

As a further example of the variation of material properties, we can consider applying hydrostatic pressure to a bulk semiconductor. This reduces the bond length and hence increases the strength of the covalent bonding interaction between neighbouring sites, but will have little effect on the bond ionicity. We would, therefore, expect the semiconductor band gaps to increase smoothly with pressure. This is indeed observed experimentally, with the direct energy gap at the Γ point typically increasing by $10\,meV$ per kilobar of applied hydrostatic pressure in III–V semiconductors.

We can estimate the bond length dependence of covalent interactions using the free-electron 'empty lattice' band structure of fig. 3.14(c). The magnitude of the wavevectors k at the Brillouin zone edges scales inversely with bond length, as $1/d$ in fig. 3.14(c). As the free electron band energies vary as $\hbar^2 k^2/2m$, we, therefore, expect the zone edge (and zone centre) energies to scale as $1/d^2$. The splitting between the bonding and anti-bonding levels will then also scale as $1/d^2$, suggesting that the nearest-neighbour interatomic covalent interactions scale as $(bondlength)^{-2}$ in tetrahedrally bonded semiconductors. This assumption has underpinned the development of semi-empirical tight-binding methods which have proved remarkably successful in predicting trends in a wide range of semiconductor electronic properties (Harrison 1989; Vogl et al. 1983).

4.5   Impurities in semiconductors

We have so far considered perfect semiconductor crystals, with every lattice site in a Si crystal, for instance, occupied by a Si atom. This ideal is never achieved in practice. There will always be some impurity atoms introduced during crystal growth, either intentionally or otherwise. Some impurity atoms can be positively beneficial, introducing shallow defect levels in the band gap, close to the band edges. Such impurities can be used to 'dope' the semiconductor. Replacing a Si atom by an arsenic atom will introduce an extra electron into the conduction band, giving what is referred to as n-type conduction ('n' for negative). By contrast, replacing a Si atom by boron removes a negatively-charged electron from the filled valence band. This is equivalent to introducing a positively-charged 'hole'. Boron is referred to as a p-type dopant ('p' for positive). Atoms which contribute an extra electron are often referred to as 'donors', while those which remove an electron from the valence band are called 'acceptors'.

In addition, there are other impurity atoms such as nitrogen, which introduce defect states in the Si energy gap, well away from the band edges. These are referred to as deep levels. They can often act as non-radiative recombination centres, trapping free carriers moving through the crystal, and hence nullifying the effect of any shallow impurities present. The control of impurity dopant atoms is the key to almost all semiconductor technology. Many of the key characteristics of shallow and deep impurity levels can be understood based on the models developed in earlier chapters.

4.5.1 Shallow impurities

Consider adding an extra electron to a semiconductor. The electron is free to move, with an effective mass, m_c^*, at the bottom of the conduction band. In practice the extra electron is introduced via an impurity atom, for example, by replacing a Si atom with an arsenic atom in a Si crystal. A neutral arsenic atom has five valence electrons which can contribute to bonding. Four of the electrons can form bonds with the neighbouring Si atoms, leaving one electron free to move through the crystal. This electron will then see a net positive charge associated with the arsenic impurity atom, and will experience a potential $-e^2/4\pi\varepsilon_0\varepsilon_r r$ due to that positive charge, where ε_r is the dielectric constant of the Si crystal. It can be shown that we can write the Hamiltonian H describing the motion of the electron as

$$H\psi(r) = \left(-\frac{\hbar}{2m_c^*}\nabla^2 - \frac{e^2}{4\pi\varepsilon_0\varepsilon_r r}\right)\psi(r) = E\psi(r) \qquad (4.19)$$

that is, treating the electron as if it were a particle with mass m_c^* moving in the impurity potential, with the zero of potential in eq. (4.19) taken to be at the bottom of the conduction band.

Equation (4.19) is identical to the Hamiltonian for an electron in an isolated hydrogen atom discussed in Appendix A, if we replace the free electron mass m by the effective mass, m_c^*, and the free space permittivity ε_0 by the semiconductor permittivity $\varepsilon_0\varepsilon_r$ in eq. (A.2). The energy of the nth bound state of an isolated hydrogen atom, E_n^H, is given in Appendix B by

$$E_n^H = -\frac{me^4}{8(\varepsilon_0 hn)^2} \qquad (4.20)$$

with the ground state energy, $E_1^H = -13.6\,\text{eV}$. By analogy, we would, therefore, expect the binding energy of the nth shallow impurity bound state, E_n^{imp}, to be given by

$$E_n^{\text{imp}} = \frac{m_{\text{rc}}^*}{\varepsilon_{\text{r}}^2} E_n^H \tag{4.21}$$

where m_{rc}^* is the relative effective mass, m_c^*/m.

Consider for instance bound electron states in GaAs. From Table 4.1 the relative electron effective mass, $m_{\text{rc}}^* = 0.067$ and dielectric constant $\varepsilon_{\text{r}} = 12.8$, then the shallow impurity ground state binding energy will be of order 6 meV. As the thermal energy, kT, at room temperature equals 25 meV, we expect the impurity level to be almost certainly ionised, with the electron then free to contribute to the room temperature conductivity of the semiconductor.

The impurity ground state should become occupied at lower temperatures, as $T \to 0$. However the electron can still be relatively delocalised. The ground state wavefunction in a hydrogen atom decays away from the nucleus as $\exp(-r/a_0)$, where a_0, referred to as the Bohr radius, is given by

$$a_0 = \frac{\varepsilon_0 h^2}{\pi m e^2} \tag{4.22}$$

We should expect the wavefunction associated with the impurity ground state level to decay away from the impurity site in a similar manner, as e^{-r/a^*}, where a^* is the defect electron effective Bohr radius, related to a_0 by

$$a^* = \frac{\varepsilon_r}{m_{\text{rc}}^*} a_0 \tag{4.23}$$

As $a_0 = 0.53\,\text{Å}$, this gives $a^* \sim 100\,\text{Å}(10^{-6}\,\text{cm})$. Hence neighbouring impurity levels will interact with each other and the levels will broaden into an impurity band for doping densities, N_d, of order $1/(10^{-6})^3\,\text{cm}^{-3}$, that is, $N_d \sim 10^{18}\,\text{cm}^{-3}$. As each impurity contributes one electron to the band, the band will be half-filled, and for sufficiently high doping densities, metallic-type conduction can occur at low temperatures in this impurity band.

The study of conduction in such impurity bands has been very fruitful in a number of research areas. These include the investigation of the transition from insulating (low doping density) to metallic behaviour at higher doping densities, referred to as the metal–insulator transition (Mott 1974). Impurity bands have also been used to investigate features of the conductivity of random systems, taking advantage of the fact that the impurity atoms do not lie on a regular array of sites, and so form a random lattice.

4.5.2 Deep impurities

As well as donors and acceptors, it is also possible to find deep impurities in semiconductors. Archetypal examples include N in GaP and Se in Si, for both of which the impurity levels lie deep in the energy gap, well away from the band edges. We aim to review here some of the origins and trends in such deep impurities. We concentrate on sp^3-bonded defects, typically formed by atoms from columns III to VII in the Periodic Table. We do not consider transition metal impurities, although these also form an important class of deep levels in semiconductors. An impurity atom can alter the periodic crystal potential in three main ways, as illustrated in fig. 4.6:

1   by attracting electrons more (or less) strongly than the host atoms, that is, having a different electronegativity;
2   by introducing lattice distortions, which alter the bond length, and hence the strength of the covalent interactions; and
3   shallow impurity states in particular can add a long-range Coulomb potential about the defect site.

The first of these three, the difference in electronegativity, is the most important factor in determining trends in deep impurity levels, with local lattice distortions, however, also playing a significant role.

To understand trends in deep levels, we first consider the hypothetical diatomic molecule, HX, where the atomic energy level, E_X, of the X atom is varied with respect to the hydrogen level, E_H. Recall from Chapter 2 that in the hydrogen molecule, H_2, the atomic orbitals on neighbouring atoms interact with each other to give a bonding and anti-bonding level at $E_H + V$

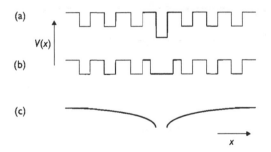

Figure 4.6   An impurity centre can alter a periodic potential by (a) introducing a deeper (or shallower) local potential; (b) distorting the nearest neighbour separation; or (c) if charged, introducing a long range Coulomb potential about the impurity centre. The potential due to the impurity centre is in each case indicated by the heavy solid line.

and at $E_H - V$, respectively, where V is negative and is a measure of the strength of the covalent interaction between the neighbouring orbitals. By comparison with the LiH molecule in Chapter 2, the two lowest energy levels in the HX molecule are given by

$$E = \frac{E_X + E_H}{2} \pm \sqrt{\left(\frac{E_X - E_H}{2}\right)^2 + V^2} \qquad (4.24)$$

The variation in the energy levels of the HX molecule with the X atom self-energy, E_X, is plotted in fig. 4.7(a). On the upper left-hand side of the

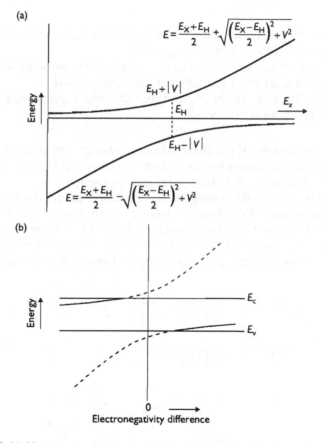

Figure 4.7 (a) Variation in the energy levels of the HX molecule as the X atom self-energy, E_X, is varied with respect to the H atom value, E_H. (b) In a semiconductor, the bonding and anti-bonding levels are broadened into bands. For small electronegativity differences, defect-related resonant states are found in the bands (dashed lines), which shift into the energy gap (solid lines) as the electronegativity difference increases.

figure, the anti-bonding level approaches the isolated hydrogen atom self-energy, E_H, as the X atom electronegativity increases (E_X decreases). On the lower right-hand side, the bonding level approaches E_H, from below, as E_X increases and the X atom becomes significantly less electronegative than the H atom. In summary, as $|E_H - E_X|$, and the electronegativity difference increases, one of the energy levels always approaches the isolated hydrogen atom self-energy, E_H.

We can now apply the same picture to a tetrahedral semiconductor, where the host atom electronegativity is fixed (as was E_H above) while the impurity electronegativity depends on the impurity considered. In semiconductors, the bonding and anti-bonding levels have broadened into bands. For sufficiently large electronegativity difference, the defect-associated state can lie in the band gap, giving a deep level, as illustrated by the thick solid lines in fig. 4.7(b), while for smaller electronegativity differences the defect states will lie in the bands, as indicated by the dashed lines. Such defect-related levels in the bands are referred to as 'resonant' levels. Because the resonant level is degenerate with the conduction or valence band, an electron will not remain bound in such a level but can instead escape into the extended band states. Hydrostatic pressure can be a very useful tool to study impurity states. The Γ conduction band edge shifts upwards with pressure, and can thereby induce transitions from resonant to deep states, as for example with Si donors in GaAs. Such transitions can also be observed in semiconductor alloys as a function of alloy composition: Si is a resonant state in $Al_xGa_{1-x}As$ for $x < 0.2$, but becomes deep at higher aluminium compositions.

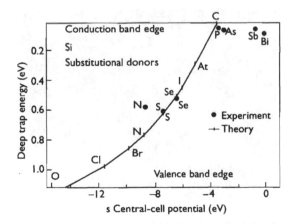

Figure 4.8 Comparison of calculated and experimentally determined defect energy levels for a range of substitutional donors in Si (figure after Vogl, © 1984 by Academic Press, reproduced by permission).

Figure 4.8 shows as an example the good agreement which can be obtained using this model between calculated energies and experimental data for a range of substitutional donors in Si. The theory omits the long range Coulomb potential, hence all the shallow impurities have zero binding energy.

4.6   Amorphous semiconductors

So far we have considered only crystalline solids. It is also possible to form amorphous semiconductors, in which each atom has a similar, covalently-bonded, local environment but there is no long range order. Figure 4.9 shows a computer-generated model of amorphous silicon (a-Si), where each atom is tetrahedrally bonded, as in the crystalline case, but no two atoms shown have exactly the same longer range environment.

In the absence of long range order, the concept of a unit cell becomes less clear-cut. Bloch's theorem no longer holds. We cannot then make the simplification of associating a wavevector k with each wavefunction or even assume that each wavefunction is spread uniformly throughout the solid.

Should we expect a band gap in a-Si? We used Bloch's theorem earlier to introduce the concept of a band gap, or forbidden energy range, in a solid. The existence of a band gap does not however depend completely on having a periodic structure. We saw in Chapter 2 that Si forms tetrahedral bonds because of the energy gained due to the strong interactions between sp^3 orbitals on neighbouring atoms. These interactions lead to widely separated bonding and anti-bonding levels, which are then

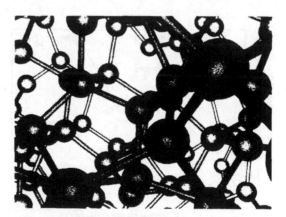

Figure 4.9 Computer-generated model of a-Si : H each atom is tetrahedrally bonded, but (unlike in crystalline Si) there is no long range order. Courtesy of M. F. Thorpe and Ming Lei – unpublished using the coordinates from Djordjevic, Thorpe and Wooten (1995).

broadened by further interactions into the valence and conduction band, respectively. The same model can clearly be applied to a-Si and it is indeed found that a-Si has an energy gap of about 1.8 eV separating the valence and conduction band. However, whereas Bloch's theorem implies that the forbidden energy range is very well defined in a crystalline solid, this is no longer true in the amorphous case. Because different atoms have slightly different local environments, including small variations in the nearest neighbour bond lengths and bond angles, the band edge energy can now vary through the solid. Figure 4.10 compares the situation in a crystalline and amorphous semiconductor: the density of allowed energy states goes to zero at a specific band edge energy in the crystalline case, but decreases approximately exponentially into the band gap in the amorphous case.

The states in the exponential tail can be regarded as defect levels, with each level localised in a particular region of the solid. Well away from the band edges, deep into the bands, we expect that the states can be extended throughout the solid, as in the crystalline case. Considerable effort has been devoted to studying the transition from localised to extended states in disordered systems. It has been shown that this transition takes place at a well-defined energy (the 'localisation edge', E_L, in fig. 4.10(b)), with the character of the localised states and localisation transition also depending on whether conduction is taking place in a three-, two-, or one-dimensional system (Elliott 1990).

The computer-generated model of a-Si in fig. 4.9 is highly idealised, with each Si atom having four Si neighbours. In practice, a-Si contains a high density of defects, with about 1 in 10^3 Si atoms having only three Si

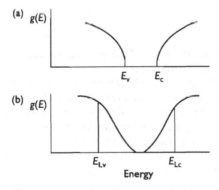

Figure 4.10 Density of allowed energy states, $g(E)$, in (a) a crystalline semiconductor, where the band edges (E_v, E_c) and energy gap are well defined, compared to (b) an amorphous semiconductor, where the density of states decreases approximately exponentially into the energy gap. The states between E_{Lv} and E_{Lc} are localised; below and above these 'localisation' or 'mobility' edges, the electronic states extend through the solid.

neighbours, which leaves one unbonded sp^3 'dangling bond' on the defect sites. This leads to a large density of deep defect states in the band gap, so that pure a-Si has little practical use.

The high density of deep defect levels can be eliminated by forming a-Si in a hydrogen-rich atmosphere, so that hydrogen atoms bond to the dangling bonds, eliminating the non-bonding defect levels in the band gap. Hydrogenated a-Si (a-Si : H) can also be doped both p- and n-type using boron and phosphorus, respectively. The carrier mobility and overall conductivity are substantially lower than in crystalline Si, both because of the disorder and also because of a lower doping efficiency. Nevertheless, a-Si : H has several important applications, chiefly because thin films of the material can be grown relatively cheaply over large areas, making it suitable both for solar cell manufacture and also to provide the switching elements for large-area, flat screen, liquid crystal displays.

In summary, we have described here some of the key features related to the electronic band structure of bulk semiconductors. We introduced the concept of effective mass and also the concept of positively-charged holes. We showed how the $k \cdot p$ method can be used to understand common features and trends in the electronic properties, including the effects of alloying and hydrostatic pressure. We also considered the effects of impurity atoms and what happens when the crystalline periodicity breaks down, giving an amorphous semiconductor. Several of these results will be used further in the next chapter when we turn to consider band structure engineering and its applications in low-dimensional semiconductor structures.

References

Bastard, G. (1988) *Wave Mechanics Applied to Semiconductor Heterostructures*, Les Editions de Physique, Courtaboeuf.

Casey, H. H. and M. B. Pannish (1978) *Heterostructure Lasers, Part B, Materials and Operating Characteristics*, Academic Press, New York.

Chelikowsky, J. R. and M. L. Cohen (1976) *Phys. Rev. B* **14** 556.

Djordjević, B. J., M. F. Thorpe and F. Wooten (1995) *Phys. Rev. B* **52** 5685.

Elliott, S. R. (1990) *Physics of Amorphous Materials*, 2nd edn, Wiley, Chichester.

Harrison, W. A. (1989) *Electronic Structure and the Properties of Solids: The Physics of the Chemical Bond*, Dover Publications, New York.

Kane, E. O. (1966) in *Physics of III–V Compounds*, Vol. 1, ed. R. K. Willardson and A. C. Beer, Academic Press, New York.

Krijn, M. P. C. M. (1991) *Semicond. Sci. Technol.* **6** 27.

Madelung, O. (ed.) (1982) *Numerical Data and Functional Relationships in Science and Technology*, (Landolt-Bornstein, New Series, Group III), vol. 17a, Springer-Verlag, Heidelberg.

Meney, A. T., E. P. O'Reilly and A. R. Adams (1996) *Semicond. Sci. Technol.* **11** 897.

Mott, N. F. (1974) *Metal–Insulator Transitions*, Taylor and Francis, London.

Vogl, P., H. J. Hjalmarson and J. D. Dow (1983) *J. Phys. Chem. Solids* **44** 365.

Vogl, P. (1984) *Advances in Electronics and Electron Physics* **62** 101.

Wei, S. H. and A. Zunger, (1996) *Appl. Phys. Lett.* **69** 2719.

Problems

4.1 Use eq. (4.9a) to estimate the value of the energy parameter E_p for the direct gap semiconductors listed in Table 4.1. Can you observe any trends in the calculated E_p values?

4.2 Use eq. (4.9b) and the values of E_p calculated in problem 4.1 to estimate the light-hole relative effective masses of the direct gap semiconductors listed in Table 4.1. Compare the calculated masses with the experimentally determined values listed in the table.

4.3 Use the data in Table 4.1 to calculate the donor ground state binding energy, E_1^{imp}, and defect electron effective Bohr radius, a^*, in InSb, InP and GaN. Hence estimate the donor doping density N_d required to achieve impurity band conduction at low temperatures in each of these compounds.

4.4 We use the $\mathbf{k} \cdot \mathbf{p}$ method in Section 4.2 and Appendix E to determine the band structure in the neighbourhood of $\mathbf{k} = 0$ using the zone centre ($\mathbf{k} = 0$) wavefunctions and energies, and the $\mathbf{k}$-dependent perturbation Hamiltonian of eq. (4.3). Show that we can generalise the $\mathbf{k} \cdot \mathbf{p}$ method to determine the band structure in the neighbourhood of an arbitrary wavevector $\mathbf{k}_0$ by introducing the wavevector $\mathbf{q} = \mathbf{k} - \mathbf{k}_0$ and then re-arranging Schrödinger's equation so that the $\mathbf{q}$-dependent Hamiltonian $H_q = e^{-i\mathbf{q} \cdot \mathbf{r}} H_0 e^{i\mathbf{q} \cdot \mathbf{r}}$ acts on the $\mathbf{q}$-independent wavefunction, $\exp(i\mathbf{k}_0 \cdot \mathbf{r}) u_{n\mathbf{k}}(\mathbf{r})$. Show that the dispersion along direction i near the state at energy $E_n(\mathbf{k}_0)$ is then given in second order perturbation theory by

$$E_n(k) = E_n(k_0) + \frac{\hbar^2 q^2}{2m} + \frac{\hbar^2 q^2}{m^2} \sum_{n' \neq n} \frac{|i \cdot p'_{nn'}|^2}{E_n(k_0) - E_{n'}(k_0)}$$

4.5 Consider the Kronig–Penney (K–P) potential of fig. 3.8 with period L and a thin barrier of height V_0 and width b centred on the origin. We saw in Chapter 3 that the NFE wavefunction for the lower state at the nth energy gap is given by $\psi_{Ln}(x) = (2/L)^{1/2} \sin(n\pi x/L)$, while the NFE wavefunction for the upper state is given by $\psi_{Un}(x) = (2/L)^{1/2} \cos(n\pi x/L)$. Show that the magnitude of the momentum matrix element P_{LUn} linking the nth upper and lower state varies as the inverse of the lattice constant L and is given by

$$P_{LUn} = i\hbar\, n\pi/L.$$

4.6   Use the $\mathbf{k} \cdot \mathbf{p}$ method and the matrix element P_{LUn} to determine the band dispersion of the upper and lower band near the nth energy gap in the K–P potential of problem 4.5. Compare the calculated $\mathbf{k} \cdot \mathbf{p}$ dispersion with that obtained using the NFE method in Chapter 3.

4.7   Consider the lowest zone centre energy gap for the K–P potential of problem 4.6, for which the NFE wavefunctions of the upper and lower states are given by $\psi_U(x) = (2/L)^{1/2} \cos(2\pi x/L)$ and $\psi_L(x) = (2/L)^{1/2} \sin(2\pi x/L)$ respectively. Use the $\mathbf{k} \cdot \mathbf{p}$ method and perturbation theory to calculate the first order correction to the wavefunction for wavevectors k near the Brillouin zone centre ($k = 0$). Hence calculate the average potential energy, U, of the upper and lower states at wavevector k close to the zone centre. Show that U *decreases* with k for the upper state, and *increases* with k for the lower state. Knowing the total energy $E(k)$ of each state from problem 4.8, and that the total energy is the sum of the kinetic and potential energy, $E(k) = T(k) + U(k)$, calculate the variation of the kinetic energy $T(k)$ for the upper and lower bands, respectively, showing that the upper band has a positive and the lower band a negative effective mass.

4.8   Consider a filled band in a 1-D crystal with lattice constant L. An electron with wavevector k and energy E when subject to an externally applied force $F = -eE$ experiences an acceleration a given by

$$a = \frac{1}{\hbar^2}\frac{d^2E}{dk^2}F \tag{4.16}$$

By integrating over all states in the filled band, show that the average electron acceleration is zero, and hence that a filled band does not contribute to the current flow in a periodic solid.

Chapter 5

Physics and applications of low-dimensional semiconductor structures

5.1 Introduction

Much of the recent interest in semiconductor materials has focussed on the ability to alter the composition and hence the band structure on the atomic scale, layer by layer. This has opened the possibility of 'band structure engineering' and even 'wavefunction engineering', whereby the composition and hence electronic and optoelectronic properties can be tailored and optimised for specific applications, introducing band structure features which would be impossible to achieve in any bulk semiconductor.

The development of molecular beam epitaxy (MBE) and metal-organic vapour phase epitaxy (MOVPE) has made it possible to produce semi-conductor 'quantum well' structures, where the composition changes on the scale of an atomic layer, with the band edge energies and carrier effective masses, consequently, changing on the same length scale. Figure 5.1 shows a high-resolution transmission electron micrograph (TEM) of several periods of a gallium indium arsenide–aluminium indium arsenide (GaInAs–AlInAs) quantum cascade laser structure, which demonstrates the growth control which can now be achieved. The white lines represent layers of GaInAs and the black lines layers of AlInAs. We shall discuss the workings of such a laser at the end of this chapter but emphasise here that the thinnest layers in this device structure are about 1 nm (three atomic layers) wide. The picture illustrates vividly the possibility of using epitaxial growth techniques, where atoms are deposited layer-by-layer to reproducibly create uniform, thin layers, controlling the composition over a very short length scale.

Such sharp changes in composition allow the energy gap to change abruptly from one layer to the next. Figure 5.2 illustrates schematically the variation of the conduction and valence band edge energies, E_c and E_v, through three layers of such a structure. The allowed energy regions for both the electrons and holes form a square well potential, similar to that considered earlier in the review of quantum mechanics in Chapter 1.

Figure 5.1 High resolution transmission electron micrograph (TEM) of a GaInAs/AlInAs quantum cascade laser structure grown by molecular beam epitaxy. The TEM contrast is set so that GaInAs and AlInAs layers appear in a light and in a dark shade, respectively. The picture illustrates both the layer reproducibility which can be achieved with epitaxial growth techniques and also the ability to vary the composition atomic layer by atomic layer. (Picture courtesy of Claire Gmachl and Federico Capasso, Bell Laboratories.)

For well widths of order 100 Å or less, both the electrons and holes experience significant quantum confinement effects. The energy levels are quantised along the growth direction. The carriers are then confined in this dimension, although they remain free to move in the other two dimensions, in the plane perpendicular to the growth axis.

The motion may in principle be further restricted to one dimension by ultrafine lithography on quantum well structures or by the growth of the alloy in V-shaped grooves, as illustrated in fig. 5.3(a) (Biasiol and Kapon 1998). If the carriers are further restricted to a quantum box or dot (see fig. 5.3(b)), the allowed energy levels and density of states resemble those

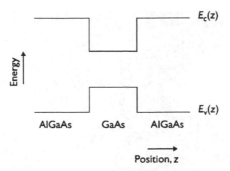

Figure 5.2 Variation of the conduction and valence band edge energies, $E_c(z)$ and
$E_v(z)$ along the growth direction through three layers of a GaAs/AlGaAs
multilayer structure.

for electrons in a large isolated atom. At each stage the likelihood of an electron or hole occupying an energy state other than the lowest is decreased, so allowing the possibility of more efficient semiconductor lasers. As a further example, the ability to physically separate dopant atoms from the layers where conduction occurs, so-called modulation doping, can lead to faster electronic devices, with improved transport characteristics. New physical effects are also found in low-dimensional structures, including the quantum Hall effect and fractional quantum Hall effect.

We can describe many of these effects using effective mass theory, where we presume the carriers behave as free particles with an effective mass, m^*, as we did when considering shallow impurity levels in Section 4.5. We begin first by considering carrier confinement and the dependence of the electron density of states on the dimensionality of a system.

5.2 Confined states in quantum wells, wires, and dots

A layered structure grown along the z-direction, such as that in fig. 5.1, remains periodic in the x–y plane. Bloch's theorem still holds in two dimensions, and we can describe the energy states in terms of the two-dimensional wavevector (k_x, k_y), as $E = E(k_x, k_y)$. If we consider the states at the centre of the two-dimensional Brillouin zone ($k_x = k_y = 0$), the band edge mismatch implies that both the electrons and holes see a quantum well potential. In the effective mass approximation, the allowed energy levels for the electrons are then found by solving Schrödinger's equation:

$$\left[-\frac{\hbar^2}{2m_c^*} \frac{\mathrm{d}^2}{\mathrm{d}z^2} + V(z) \right] \psi(z) = E\psi(z) \tag{5.1}$$

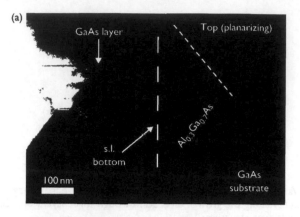

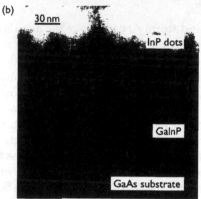

Figure 5.3 (a) TEM micrograph of a quantum wire structure, achieved by growing a multilayer structure in a V-shaped groove. Variation of alloy growth rate on the side walls and base of the groove lead to the formation of quantum wires at the bottom of the groove, in which carriers are free only to move along the direction of the groove (from Biasiol and Kapon, © 1998 by the American Physical Society). (b) TEM of a quantum dot structure: when, for example, InP is grown on GaInP under appropriate growth conditions, the lattice-mismatch between the InP and GaInP leads to the formation of 'self-ordered' InP islands in the GaInP matrix (from Eberl 1997).

where $\psi(z)$ is the electron envelope function and $V(z)$ is the band edge potential distribution, given in the quantum well case of fig. 5.2 by $V(z) = E_c(z)$. To calculate the electron (and hole) states in such a square well, we require at the interface between the well, W, and barrier, B, layers that

$$\psi_W = \psi_B \tag{5.2a}$$

and

$$\frac{1}{m_W^*} \frac{d\psi_W}{dz} = \frac{1}{m_B^*} \frac{d\psi_B}{dz} \tag{5.2b}$$

This second relation is a generalisation of the earlier condition in Chapter 1, that $d\psi_W/dz = d\psi_B/dz$, to the case where the mass changes on crossing the boundary. Equation (5.2b) ensures what is referred to as the 'conservation of probability current density' between different layers (Bastard 1988). The confinement energies are then found by solving

$$k \tan\left(\frac{kL}{2}\right) = \frac{m_W^*}{m_B^*} \kappa \tag{5.3a}$$

for states of even parity, while for states of odd parity

$$-k \cot\left(\frac{kL}{2}\right) = \frac{m_W^*}{m_B^*} \kappa \tag{5.3b}$$

where L is the well width, with $k^2 = 2m_W^* E/\hbar^2$ and $\kappa^2 = 2m_B^*(\Delta E_c - E)/\hbar^2$. ΔE_c is the conduction band offset, and the zero of energy is at the bottom of the well.

We can extend the analysis to quantum wires and dots. If we assume an infinite confining potential ($V = \infty$ in barrier) and rectangular wires and dots, then the allowed energy states of the confined electrons are given by

$$E_i(k_x, k_y) = \frac{\hbar^2 i^2}{8m^* L_z^2} + \frac{\hbar^2}{2m^*}(k_x^2 + k_y^2) \qquad \text{Quantum well}$$

$$E_{i,j}(k_x) = \frac{\hbar^2}{8m^*}\left(\frac{i^2}{L_z^2} + \frac{j^2}{L_y^2}\right) + \frac{\hbar^2 k_x^2}{2m^*} \qquad \text{Quantum wire} \tag{5.4}$$

$$E_{i,j,k} = \frac{\hbar^2}{8m^*}\left(\frac{i^2}{L_z^2} + \frac{j^2}{L_y^2} + \frac{k^2}{L_x^2}\right) \qquad \text{Quantum box}$$

where L_z, L_y, and L_x are the confining dimensions, $i, j, k = 1, 2, \ldots$ are the quantum confinement numbers, k_x, k_y are the wavevector components along the unconfined directions, m^* is the carrier effective mass, and we assume the zero of energy at the confined layer conduction band edge.

5.3 Density of states in quantum wells, wires, and dots

Limiting the electron motion to fewer dimensions dramatically modifies the electron energy spectrum, leading to an enhancement of the density of states near the band edge. To see this, we need to calculate the electronic density of states for a D-dimensional crystal, whose sides are of length L, and in which the dispersion near a band edge is given by the parabolic dispersion,

$$E = \frac{\hbar^2 k^2}{2m^*} \tag{5.5}$$

Not all values of k are allowed in eq. (5.5). We require that the allowed solutions of Schrödinger's equation satisfy the boundary conditions appropriate to the given potential. For a given crystal, we therefore require that the wavefunctions decay to zero at the crystal surfaces. The existence of the surface implies that, strictly speaking, Bloch's theorem should not be applied within a finite crystal. But we know that Bloch's theorem works and is very useful for describing many crystal properties. We would like to maintain Bloch's theorem, while still recognising that we have a finite crystal of size L. We can do so by introducing periodic boundary conditions on the crystal, requiring that the amplitude and derivative of each wavefunction are equal at $x = 0$ and $x = L$, so that

$$e^{ik_x L} = 1 \tag{5.6}$$

with the allowed values of k_x then given by

$$k_x L = 2\pi n \tag{5.7a}$$

or

$$k_x = 2\pi n / L \tag{5.7b}$$

where n is an integer. Likewise, in a cube of side L, we have $k_y = 2\pi p / L$ and $k_z = 2\pi q / L$, where p and q are both integers.

We can then define the density of states function, $n(E)$, such that the number of allowed energy states, dN, between energy E and $E + dE$ is given by

$$dN = n(E)dE \tag{5.8}$$

The number of allowed states in this energy range will depend directly on the number of states whose allowed wavevectors fall in this range, that is,

$$n(E)dE = n(k)dk \tag{5.9}$$

where $n(k)$ is the density of allowed k points.

Re-arranging eq. (5.9), the density of electronic states, $n(E)$, is then given by

$$n(E) = n(k)\frac{dk}{dE} = \frac{n(k)}{dE/dk} \tag{5.10}$$

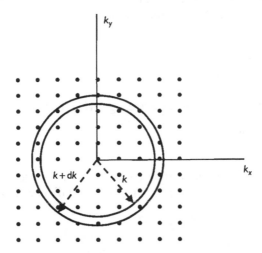

Figure 5.4 The grid of allowed k-points $(2\pi n/L, 2\pi p/L)$ form a reciprocal lattice associated with a 2-D crystal of size L^2. The two circles of radius k and $k + dk$
describe contours of constant energy, with $E = \hbar^2 k^2/2m^*$ on the inner
circle.

This makes sense: if the energy, E, is changing rapidly with wavevector
k (dE/dk large), then there will be fewer states in a given energy range
than when dE/dk is small. From eq. (5.5), we have for a parabolic band
that $dE/dk = \hbar^2 k/m^*$, so that

$$n(E) = \frac{m^*}{\hbar^2 k} n(k) \tag{5.11}$$

As the density of k-states, $n(k)$, depends on the dimensionality of the
structure, so too will the density of states, $n(E)$.

Figure 5.4 shows the grid of allowed k-points $(2\pi n/L, 2\pi p/L)$, near the
origin in two dimensions. The two circles of radius k and $k+dk$ are contours
of constant energy, with $E = \hbar^2 k^2/2m^*$ on the inner circle.

In D dimensions, each k-point occupies a volume $(2\pi/L)^D$, so that the
number of k-states per unit volume of k-space is then $1/(2\pi/L)^D = (L/2\pi)^D$.
With two allowed electron states per k-value (one spin up, and one spin
down), the density of allowed states is $2(L/2\pi)^D$.

Turning first to the two-dimensional case of fig. 5.4, the area between
the two rings of radius k and $k + dk$ equals $2\pi k\,dk$, so that the number of
allowed states between k and $k + dk$, $dN(k)$, is given by

$$dN(k) = 2(L/2\pi)^2\, 2\pi k\,dk = n_{2D}(k)dk \tag{5.12}$$

That is, $n_{2D}(k) = kL^2/\pi$. Substituting this into eq. (5.11), we find that the density of allowed electron states near a band edge in a 2-D structure of area L^2 is given by

$$n_{2D}(E) = \frac{m^*}{\pi} \left(\frac{L}{\hbar}\right)^2 \tag{5.13}$$

with the density of states per unit area, $g_{2D}(E)$, then given by

$$g_{2D}(E) = \frac{4\pi m^*}{h^2} \tag{5.14}$$

We can go through a similar analysis in both the 1-D and 3-D cases. The density of states per unit volume, $g_{3D}(E)$, in a bulk crystal is given by

$$g_{3D}(E) = 4\pi \left(\frac{2m^*}{h^2}\right)^{3/2} E^{1/2} \tag{5.15}$$

while in 1-D, the density of states per unit length, $g_{1D}(E)$ is given by

$$g_{1D}(E) = \frac{2(2m^*)^{1/2}}{h} E^{-1/2} \tag{5.16}$$

Figure 5.5 illustrates schematically the change in the density of states, $g(E)$, as the electron motion is limited to fewer dimensions in a semiconductor structure. The most striking feature observed as the dimensionality is

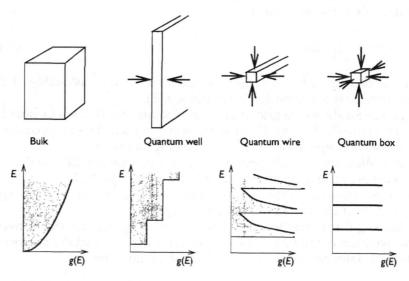

Figure 5.5 The change in the density of states, $g(E)$, as the number of confining dimensions is increased.

reduced is the build-up in the density of states near the band edge. This has significant potential benefits for semiconductor lasers: a greater proportion of the injected carriers can then in principle be in states contributing to band edge population inversion and gain. The development of lasers containing a uniform array of quantum boxes or quantum wires presents considerable growth and fabrication problems; however, quantum well lasers are well established and widely available commercially.

5.4   Modulation doping and heterojunctions

The conductivity, σ, in a bulk semiconductor, depends on the carrier density per unit volume, n, and the carrier mobility, μ, as

$$\sigma = ne\mu \tag{5.17}$$

The obvious route to increasing conductivity, then is, to increase the carrier density by increasing the doping density, N_d. However, this also increases the number of ionised impurity scattering centres $(= N_d)$, thereby reducing the mobility, particularly at lower temperatures.

By contrast, the areal carrier density, n_s, can be increased in a low-dimensional system without significantly degrading the mobility. This can be achieved through modulation doping, where the dopant atoms are placed in a different layer to that in which conduction is occurring. This is illustrated in fig. 5.6a, where donor atoms are placed in the barrier layers adjacent to a quantum well. The excess donor electrons are transferred

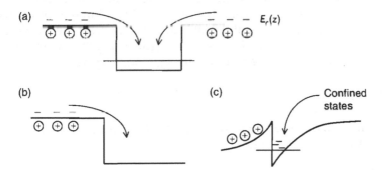

Figure 5.6 (a) When donor atoms are placed in the barrier layers adjacent to a quantum well, the excess electrons $(-)$ can transfer into the quantum well, leaving the positively charged ionised impurity centres $(\oplus)$ in the barrier. (b) A modulation-doped heterojunction can be formed by doping a thin region of a wide-gap semiconductor close to the interface with a narrower gap material. (c) It is then energetically favourable for the electrons to transfer into the narrower gap material. The electrons become confined at the heterojunction because of the electrostatic potential due to the positively charged ionised impurity sites $(\oplus)$.

into the quantum well, leaving the ionised impurity centres in the barrier, typically over 100 Å from the well.

It is also possible by modulation doping to achieve 2-D conduction at a heterojunction between two layers of differing band gap. Consider, for instance, doping a thin layer of AlGaAs with an areal doping density, N_s, with the doping layer separated by an undoped spacer layer of width w from a neighbouring GaAs region (fig. 5.6b).

It is energetically favourable to transfer the doping electrons into the narrower band gap GaAs layer. If n_s electrons are transferred across per unit area, this will leave a fixed positive charge associated with the ionised impurity atoms, and induce a built-in electric field, E, in the spacer layer, of magnitude

$$E = \frac{en_s}{\varepsilon_0\varepsilon_r} \tag{5.18}$$

with a consequent linear variation in potential across the spacer layer (fig. 5.6c).

The heterojunction potential and confined state energies should be determined self-consistently, as the confined electron states, and their wavefunctions, $\psi(z)$, will depend on the potential $V(z)$, while the potential variation is in turn determined by the electron spatial distribution (proportional to $\sum |\psi(z)|^2$) at the heterojunction.

It is beyond our scope to calculate these confined state energies self-consistently, but we can get a qualitative understanding of the behaviour of carriers at a heterojunction by approximating to the potential and using the variational method introduced in Chapter 1 and Appendix A.

We presume the electrons are confined wholly within the narrow band-gap layer, and so set $V = \infty$ at the interface ($z = 0$). Near the interface, the electric field due to the fixed charge, eq. (5.18), is largely unscreened, and so we let the conduction band edge energy, $V(z)$, vary as

$$V(z) = \frac{e^2 n_s z}{\varepsilon_0\varepsilon_r} \qquad z > 0 \tag{5.19}$$

The trial wavefunction, $f(z)$, must satisfy the conditions $f(0) = 0$, and $f(z) \to 0$ as $z \to \infty$. We choose

$$f(z) = 0 \qquad\qquad z \leq 0$$

$$= \sqrt{\frac{b^3}{2}}\, z\, e^{-bz/2} \qquad z > 0 \tag{5.20}$$

where the normalisation constant, $(b^3/2)^{1/2}$ is chosen so that

$$\int_{-\infty}^{\infty} dz\, f^*(z) f(z) = 1 \tag{5.21}$$

This is referred to as the Fang–Howard wavefunction (Fang and Howard 1966; Bastard 1988).

To estimate the lowest confined state energy $E_1(b)$, we need first to find the value of b for which the total energy of the system per electron, $\mathcal{E}_1(b)$, is minimised. We must be careful to avoid double-counting the electron–electron interactions when evaluating $\mathcal{E}_1(b)$.

The ground state total energy per electron, $\mathcal{E}_1(b)$, is found by minimising

$$\mathcal{E}_1(b) = \frac{b^3}{2} \int_0^\infty dz\, z e^{-bz/2} \left(-\frac{\hbar^2}{2m^*} \frac{d^2}{dz^2} + \frac{1}{2} \frac{e^2 n_s z}{\varepsilon_0 \varepsilon_r} \right) z e^{-bz/2} \tag{5.22}$$

The factor $\frac{1}{2}$ in the potential term here follows by realising that while the potential experienced by a charge q in a capacitor is, say, $q\phi$, the total work done and, therefore, average energy per electron in building that potential is $\frac{1}{2}q\phi$. Evaluating eq. (5.22), we find that

$$\mathcal{E}_1(b) = \frac{\hbar^2 b^2}{8m^*} + \frac{1}{2} 3e^2 n_s over \varepsilon_0 \varepsilon_r b \tag{5.23}$$

which is minimised when

$$b_{\min} = \left(\frac{6m^* e^2 n_s}{\hbar^2 \varepsilon_0 \varepsilon_r} \right)^{1/3} \tag{5.24}$$

The one-electron confined state energies, $E_1(b_{\min})$ are then found by evaluating eq. (5.22) with $\frac{1}{2}V(z)$ replaced by $V(z)$ in the potential energy term. This gives

$$E_1 = \left(\frac{6^{2/3}}{8} + \frac{3}{6^{1/3}} \right) \left(\frac{e^2 n_s}{\varepsilon_0 \varepsilon_r} \right)^{2/3} \left(\frac{\hbar^2}{m^*} \right)^{1/3} \tag{5.25}$$

We see from eq. (5.25) that the confined state energies then depend on the carrier density per unit area, that is, the areal carrier density, n_s, as $n_s^{2/3}$. We therefore expect the separation between the first two confined states, E_1 and E_2 to vary as

$$E_2 - E_1 \sim n_s^{2/3} \tag{5.26}$$

Because the carriers at the heterojunction are still free to move in two dimensions, the density of states will be constant, and so the Fermi energy E_F will increase linearly with respect to the band minimum energy, E_1 as

$$E_F - E_1 \sim n_s \tag{5.27}$$

We thus expect that for sufficiently high carrier density, n_s, the Fermi level will enter the second sub-band. This is indeed observed experimentally. It can lead for instance to a drop in electron mobility at low temperatures, due to the additional scattering associated with electron transfer between the first and second sub-bands (see fig. 5.7).

At very low temperatures and in very pure materials, the electron mobility at GaAs/AlGaAs heterojunctions can exceed 10^6 cm^2/(V s), three orders of magnitude larger than in low-doped bulk material, due to the virtual elimination of ionised impurity scattering. The effect is much less marked at room temperature, where other scattering mechanisms dominate, in particular scattering by polar-optic phonons. Nevertheless, the room temperature mobility in modulation-doped heterojunction field effect transistors is typically double that of the doped GaAs used in metal-gate field-effect transistors. This has two important consequences for the performance of high-speed transistors: first, the resistances are reduced, and with them the RC time constants, so that devices of a given size are faster and second, largely because of the reduced resistance, the levels of noise generated by the device (due to scattering processes) are much reduced. The lowest noise transistors presently available are, therefore, based on modulation-doped heterojunctions, which find widespread application, for instance, in the amplifier circuits of domestic satellite dish receivers.

5.5 Quantum Hall effect

The Hall effect provides a well-established technique to determine the mobility and carrier density per unit area in bulk semiconductor samples (Hook and Hall 1991). It was, therefore, an obvious technique to apply to low-dimensional semiconductor structures. However, when such measurements were carried out at low temperatures, the results were completely unexpected (von Klitzing 1986). The measured Hall resistance was quantised in units of h/e^2, where h is Planck's constant, and e is the electron charge. As a consequence, a basic semiconductor experiment has become the standard for defining resistance and, more interestingly, has opened a wide field of fundamental research, including analogies that we do not pursue here but which can aid in the understanding of superconductivity (Kivelson et al. 1996).

We consider first the classical Hall effect in a 2-D sample, with the current, I, given by

$$I = wn_s ev \tag{5.28}$$

where n_s is the areal carrier density, v the average carrier velocity, and w is the width of the sample, as illustrated in fig. 5.8.

When a magnetic field, B, is applied perpendicular to the sample, it causes a force on each carrier, $\mathbf{F} = e(\mathbf{v} \times \mathbf{B})$, whose magnitude is then evB,

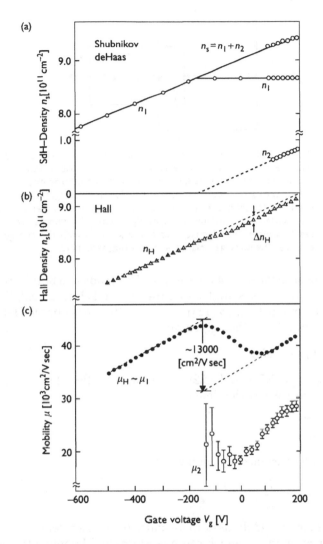

Figure 5.7 Variation in the low-temperature mobility with increasing carrier density at a heterojunction interface. The mobility decreases markedly when the carrier density increases to the point where carrier scattering becomes possible between the first and second sub-band. The carrier density was changed by varying the gate voltage across a field effect transistor structure (from Stormer, © 1982 by Elsevier Science, reproduced by permission of the publisher).

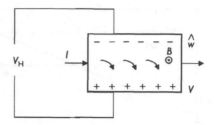

Figure 5.8 Plan view of a two-dimensional structure of width *w* with a magnetic field
B applied perpendicular to the structure (into the page). If the current flow
is due to positive charges moving from left to right, the magnetic field will
deflect the carriers towards the bottom face in the diagram, leading to a net
build-up of positive charge on this face, and negative charge on the opposing
face, giving a measurable voltage, V_H across the sample. Equilibrium is
reached when the electric field E_H associated with the charge build-up
balances the force associated with the magnetic field.

directed towards the side of the sample. This leads to a build-up of charge
on the two sides of the sample, until the induced electric field, E_H, exactly
balances the magnetic force, $eE_H = evB$, with a measurable Hall voltage,
V_H, across the sample then given by

$$V_H = E_H w = vBw \qquad (5.29)$$

Combining eqs (5.28) and (5.29), we can use the Hall voltage V_H to
determine the areal carrier density n_s, as

$$V_H = \frac{B}{n_s e} I \qquad (5.30)$$

with the Hall resistance, R_H, defined as

$$R_H = \frac{V_H}{I} = \frac{B}{n_s e} \qquad (5.31)$$

The Hall effect is widely used to measure the carrier density, n_s, and also
the carrier mobility, μ, which can be determined knowing the current, I,
carrier density and applied longitudinal voltage, V.

How then does the Hall effect become quantised in two dimensions? We
recall from eq. (5.4) that the energy levels in the ground state sub-band of
a 2-D electron gas (2DEG) satisfy the relation

$$E(k_x, k_y) = E_0 + \frac{\hbar^2}{2m^*}(k_x^2 + k_y^2) \qquad (5.4)$$

where E_0 is the ground state zone centre confinement energy and the
electrons are free to move in the *x–y* plane.

When a strong magnetic field, B, is applied perpendicular to the 2DEG, the electrons move in cyclotron orbits in the 2-D plane. Classically, the centripetal force, F, on each electron is given by

$$F = evB = \frac{m^* v^2}{r} \qquad (5.32)$$

where r is the radius of the cyclotron orbit. The cyclotron frequency, ω_c, then depends directly on the applied field, B, as

$$\omega_c = v/r = eB/m^* \qquad (5.33)$$

Classically, all values of the orbital radius, r, and electron energy, E, are allowed. However, when quantisation effects are taken into account, it can be shown that the only allowed orbital energies depend directly on ω_c as $E_n = (n + \frac{1}{2})\hbar\omega_c$, where n is an integer and the quantised energy levels are referred to as Landau levels. The energy levels of the 2DEG are then given by

$$E_n = E_0 + (n + 1/2)\hbar\omega_c + g\mu_B B \cdot s \qquad (5.34)$$

where the last term describes the interaction between the electron spin and the applied magnetic field, and is described in more detail in Chapter 6.

The form of the density of states then changes in an applied magnetic field from the constant density of states of fig. 5.5 to a series of discrete allowed energy levels, as illustrated in fig. 5.9. The total number of electron states is, however, conserved per unit energy range. We saw earlier that in zero field, the total number of states, N, per unit area between energy E and $E + dE$ is given by $N = g_{2D}(E)dE = 4\pi m^*/h^2\, dE$. All the states lying within an energy range $dE = \hbar\omega_c$ are gathered into each pair of spin up and spin down Landau levels. The number of states, N, in each individual Landau level is then given by

$$N = \frac{1}{2}\left(\frac{4\pi m^*}{h^2}\right)\hbar\omega_c = \frac{eB}{h} \qquad (5.35)$$

When j Landau levels are fully occupied, the areal carrier density $n_s = Nj$, and the Hall resistance is given by

$$R_H = \frac{B}{n_s e} = \frac{h}{je^2} \qquad (5.36)$$

It can be shown that the resistance, R_I, to the applied voltage should go to zero when the Landau levels are fully occupied and $\hbar\omega_c \gg k_B T$ (i.e. at high fields and low temperatures). $R_I = 0$ in this case, first, because the electrons cannot scatter to other (empty) states with the same energy, and,

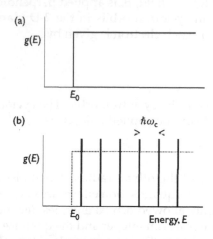

Figure 5.9 When a magnetic field B is applied perpendicular to a two-dimensional structure the form of the density of states changes from a continuous spectrum (a) to a series of discrete allowed energy levels (b), due to quantization associated with the circular motion of the electrons in the plane perpendicular to the applied magnetic field. For simplicity, the electron spin energy, $g\mu_B\boldsymbol{B} \cdot \boldsymbol{s}$ is ignored in (b).

second, because when $\hbar\omega_c \gg k_B T$ the electrons will not be scattered to other Landau levels. Given n_s carriers per unit area, we then expect that $R_I = 0$ when $n_s = Nj = jeB/h$.

In practice, it is found for many samples that $R_I = 0$, and the Hall resistance is quantised at $R_H = h/je^2$ over a finite range of field in the neighbourhood of $B = hn_s/je$, as illustrated in fig. 5.10.

The step heights in the quantum Hall effect can be measured to better than one part in 10^6 and lead to an extremely accurate determination of $h/e^2 = 25\,813\,\Omega$. A basic semiconductor experiment can, therefore, be used in defining fundamental constants (h or e), and also as a resistance standard, to define the ohm.

The model of the quantum Hall effect here is greatly oversimplified. It does not, for example, account for the width of the plateaux in R_H in fig. 5.10. The plateaux width can be explained in terms of broadening of the Landau levels, for example, by impurities and the localisation of electron states in the wings of the broadened Landau levels, as illustrated in fig. 5.11. Conduction occurs through the extended states and so, when the Fermi energy lies well within the band of localised states, the conduction electrons again see no states close in energy to which they can scatter.

Once the magnetic field is sufficiently large that all electrons are in the lowest Landau level, there should be no further plateaux in the Hall

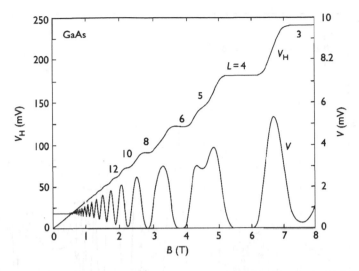

Figure 5.10 Experimental curves for the longitudinal voltage (V) and the Hall voltage (V_H) of a heterostructure as a function of the magnetic field B for a fixed carrier density in the heterostructure (after Cage et al. © 1984 IEEE).

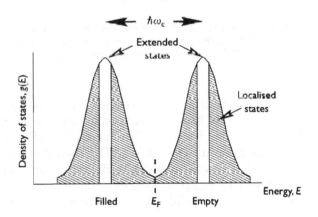

Figure 5.11 Model for the broadened density of states of a 2-D electron gas in a strong magnetic field. Mobility edges close to the centre of the Landau levels separate extended states from localised states.

resistance, R_H or zeros in R_I. It was, therefore, a further big surprise when plateaux and zeros were seen when the lowest level was one-third and two-thirds full, and then, as the material quality improved at further fractions such as $\frac{1}{5}, \frac{2}{5}, \frac{2}{7}, \frac{2}{9}$ etc. (see fig. 5.12). The theory for these plateaux

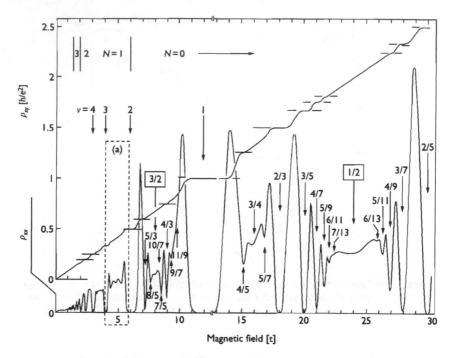

Figure 5.12 The fractional quantum Hall effect, showing evidence of a rich variety of many-body effects in a two-dimensional electron gas in a strong magnetic field (after Willett *et al.* © 1987 by the American Physical Society).

requires many-electron effects which cause energy gaps to open up within the Landau levels: there are bound states containing, for example, three electrons whose excitations have an effective charge of $\frac{1}{3}$, and which then account for the plateaux at $\frac{1}{3}$ and $\frac{2}{3}$. The theory of the fractional quantum Hall effect has several parallels with that of superconductivity, and analogies between the two processes have significantly furthered the understanding of the quantum Hall effect (Kivelson *et al.* 1996).

5.6 Semiconductor laser action

Optical sources generally require a direct gap semiconductor, where the lowest energy state in the conduction band, E_c, is directly above the highest energy state in the valence band, E_v. This has the very important consequence that an electron which has relaxed to its lowest possible energy in the conduction band can recombine directly with a hole at the top of the valence band and emit a photon.

Direct gap semiconductors can, therefore, make efficient light sources, such as lasers, light emitting diodes, and optical amplifiers. Although Si is the mainstay of the electronics industry, accounting for the vast majority of semiconductor production, its indirect band gap (described in Chapter 4) has so far ruled it out as an optical source material. Instead, optoelectronic devices are based predominantly on direct gap III–V semiconductors such as GaAs and InP. The energy gap in such semiconductors varies from about 0.1 eV in InSbAs alloys to over 6 eV in AlN, and this, in principle, enables semiconductor direct gap optical emission from about 10 μm in the mid-infra-red right through the visible spectrum to about 20 nm in the ultra-violet. Quantum well structures play a major role throughout this full spectral range.

Even in a direct band gap semiconductor, a typical electron–hole recombination lifetime is 10^{-9} s while the typical time for an electron or hole to undergo scattering with a phonon is much shorter (10^{-13}–10^{-11} s). Thus it is possible to assume that electrons in the conduction band are in thermal equilibrium with each other at the lattice temperature T and that similarly holes in the valence band are in thermal equilibrium with each other, even when electrons and holes are *not* in equilibrium with one another. We may, therefore, describe the energy distribution of electrons and holes using Fermi–Dirac statistics. That is, the probability, f_c, of a state at an energy E_e in the conduction band being occupied by an electron is

$$f_c = \frac{1}{\exp[(E_e - F_c)/kT] + 1} \tag{5.37}$$

and the probability, f_v, of a state at energy E_h in the valence band being occupied by an electron is

$$f_v = \frac{1}{\exp[(E_h - F_v)/kT] + 1} \tag{5.38}$$

where F_c and F_v are called the *quasi*-Fermi levels for electrons and holes, respectively, and k is Boltzmann's constant.

In order to achieve population inversion at the band edge in a semiconductor, and thence optical gain, we require that the probability of stimulated emission exceeds the probability of absorption. It can be shown that this occurs when the quasi-Fermi level separation exceeds the energy gap, that is, to obtain gain we need to pump electrons and holes into the semiconductor until

$$F_c - F_v \geq E_g \tag{5.39}$$

This is known as the Bernard–Duraffourg relationship (Bernard and Duraffourg 1961). The semiconductor then exhibits optical gain in the

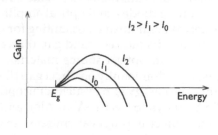

Figure 5.13 Material gain, *g*, as a function of photon energy in a semiconductor medium. Successive curves show the gain spectrum as the injected current *I* and hence the carrier density is increased. In each case, *g* = 0 at an energy $E = F_c - F_v$, the quasi-Fermi level separation for the given drive current.

energy range from E_g to $F_c - F_v$ and is absorbing at higher energies as illustrated in fig. 5.13. One of the major aims of band structure engineering in semiconductor lasers is to obtain transparency (when $F_c - F_v = E_g$) and gain for the lowest possible injected carrier and current densities.

As the current injected into the laser is increased, initially the electron and hole densities also increase and so the separation of the quasi-Fermi levels, $F_c - F_v$, increases (see fig. 5.13), leading to an increase in the peak gain value. This continues until the maximum gain is equal to the total losses from the laser cavity. Loss mechanisms within the cavity include photon scattering due to imperfections and also reabsorption mechanisms associated perhaps with defect states in the laser structure. In addition, the gain must overcome the loss of photons from the end mirrors of the laser.

The characteristics of III–V semiconductor lasers are in large part determined by the valence band structure, which is complicated even in the unstrained case (fig. 5.14a). We saw in Chapter 4 how the Light- (LH) and Heavy-Hole (HH) bands are degenerate at the zone centre, Γ, with the spin-split-off band lying at an energy E_{so} below the two highest bands. Although semiconductors are among the most efficient of laser materials there are several drawbacks associated with the band structure of fig. (5.14a). First, the conduction band effective mass is small, while the highest (HH) valence band always has a large effective mass, leading to a large valence band density of states and hence requiring a large carrier density and a large spread in electron energies at population inversion. If the conduction and valence bands had more similar shapes, a lower density of both holes and electrons, which must be equal to maintain charge neutrality, would be required. Second, in bulk lasers, cubic symmetry means that the states at the valence band maximum are made up in equal part of p_x, p_y and p_z-like

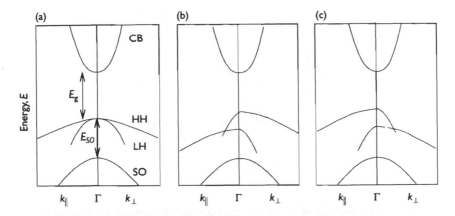

Figure 5.14 (a) Schematic representation of the band structure of an unstrained direct gap tetrahedral semiconductor. (b) Under biaxial compression, the axial strain component splits the degeneracy of the valence band maximum, and introduces an anisotropic band dispersion, with the highest band *heavy* along $k_\perp$, the strain axis (= growth direction), but *light* along $k_\parallel$ (in the growth plane). (c) Under biaxial tension, the valence band splitting is reversed, with the highest band now *light* along $k_\perp$ and comparatively *heavy* along $k_\parallel$ (from O'Reilly (1989) with permission of the Institute of Physics).

orbitals. The holes are distributed equally between the three types of states and are equally likely to produce light which is linearly polarised in the *x, y,* or *z* directions. Therefore only one-third of even those holes at the correct energy are in the right polarisation state to contribute to the lasing mode, which will be polarised along one specific direction.

The semiconductor band structure also determines the dominant loss mechanisms. At longer wavelengths (1.3 µm and beyond) Auger recombination becomes an increasingly important loss mechanism. In Auger recombination, an electron and hole recombine across the band gap but instead of emitting a photon give their energy and momentum to excite a third carrier, such as the electron in fig. 5.15. Because Auger recombination involves three carriers, the recombination current varies with carrier density, n, as Cn^3, where C is referred to as the Auger recombination coefficient. It can be shown for the case considered in fig. 5.15 that the Auger coefficient, C depends on the energy gap, E_g, and electron and hole effective masses, m_c^* and m_v^* approximately as

$$C(T) = C_0 \exp\left(-\frac{m_c^* E_g}{m_c^* + m_v^*}/kT\right) \tag{5.40}$$

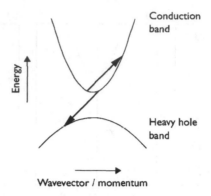

Figure 5.15 An example of an Auger recombination process in a semiconductor, where the energy and momentum released when an electron and hole recombine across the band gap is used to excite another carrier; in this case exciting an electron from near the conduction band minimum to a higher conduction state.

Auger recombination then becomes increasingly important as $E_g \rightarrow 0$, and in fact ultimately limits the room-temperature application of conventional semiconductor lasers at longer wavelengths (about 3 μm and beyond).

Several strategies are available to improve on the unstrained bulk band structure of fig. 5.14(a). The growth of quantum well, wire, or dot structures modifies the density of states, as shown in fig. 5.5. At each stage in going from bulk to quantum dot the likelihood of an electron or hole occupying an energy state other than the lowest is decreased, and the current required for lasing action is in principle reduced.

Quantum box and quantum wire lasers still present considerable growth and fabrication problems but quantum well lasers are well established and widely available commercially. In addition, the laser characteristics can be further improved by the introduction of strain in the quantum well layer.

5.7   Strained-layer lasers

It is now possible to grow high-quality, strained-layer structures, in which, for example, a single layer is composed of a semiconductor which would normally have a significantly different lattice constant to the substrate material. The lattice mismatch is accommodated by a tetragonal distortion of the layer, as illustrated in fig. 5.16, giving a built-in axial strain. The stored strain energy is linearly dependent on the thickness while a certain minimum energy is associated with the formation of a dislocation

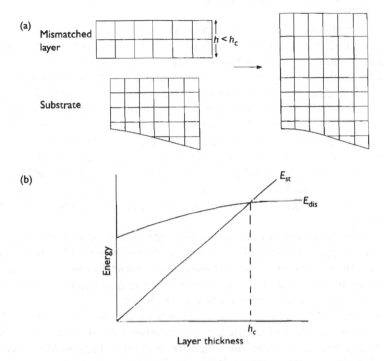

Figure 5.16 (a) Biaxial compression occurs when a layer has a larger lattice-constant than the substrate on which it is grown: the layer is compressed in the growth plane and relaxes by expanding along the growth direction. (b) Energy stored per unit area versus layer thickness h in a strained layer (E_{st}), and in a dislocation network relieving the strain (E_{dis}). The strained layer is thermodynamically stable for $h < h_c$ (from O'Reilly (1989) with permission of the Institute of Physics).

and plastic relaxation. Therefore, below a certain critical thickness, h_c, the elastically strained layers are thermodynamically stable and high quality pseudomorphic growth can be achieved.

The growth of strained-layer structures has several advantages. Strained layers allow new material combinations on established substrates, including, for instance, InGaAs alloys on GaAs or InP. With the independent variation of band gap and lattice constant, it is possible to access new band regimes not otherwise achievable, such as the InGaAs/GaAs lasers emitting at $1\,\mu m$ which are used to pump optical fibre amplifiers. The built-in strain also introduces new physical effects, significantly modifying the electronic properties of the semiconductor materials. The combination of new emission wavelengths and improved characteristics at existing wavelengths has led to widespread application of strained-layer lasers and amplifiers (Adams and O'Reilly 1992; Adams, O'Reilly and Silver 1998).

To understand the effects of strain on the band structure, we need to return to the band structure of fig. 5.14(a). The LH and HH bands are only degenerate because of the cubic symmetry of the lattice and they are split apart by strain. This splitting arises because the z-like valence state now sees a different environment to the x- and y-like states due to the change in unit cell dimensions. In addition, the resulting band structure is highly anisotropic (fig. 5.14(b) and (c)), with the band which is heavy along the strain axis, $k_\perp$, being comparatively light perpendicular to that direction, $k_\parallel$, and vice-versa. This anisotropy can be understood from the $k \cdot p$ theory introduced in Chapter 4, where we saw that the interaction between the conduction band and a z-like state gives rise to a valence band with a low effective mass along the z-direction, and heavy mass perpendicular to that direction.

When the layer is grown in compressive strain, the band with lower hole mass in the quantum well plane moves to the valence band maximum, as shown in the LHS of fig. 5.14(b). There is now a much better match between the carrier effective masses in the conduction and valence bands and the threshold current and carrier density are reduced.

The band splitting is reversed in layers under tensile strain (LHS of fig. 5.14(c)). A marked reduction in threshold current has also been observed in tensile-strained lasers. A different mechanism must be invoked to explain these improvements. We have already mentioned how the holes in a bulk laser are distributed equally between states that would produce light polarised in the x, y, or z directions, so that only one in three holes contribute to the lasing mode. However, the reduced symmetry in the tensile-strained layer shifts the z-like valence states up in energy compared to the x- and y-like states; for sufficient tensile strain, most of the holes then have z-like character and are in the right polarisation state to contribute to the lasing mode, which is indeed found to have its electric field polarised along the growth (TM) direction.

Figure 5.17 shows a compilation by Thijs et al. (1994) of the measured strain dependence of threshold current density per well in long wavelength (1.5 μm) InGaAs(P) quantum well lasers. The reduction in threshold current with both tensile (LHS) and compressive strain (RHS) is clearly seen. The maximum threshold current occurs in layers with a small tensile mismatch, rather than in lattice-matched, unstrained quantum wells. This occurs because strain and quantum confinement each split the HH and LH states at the valence band maximum. The splittings add to each other for compressive strain but are in opposite directions for tensile strain: the largest threshold current density is then found about the point where the strain- and confinement-induced splittings are equal and opposite, leaving the HH and LH bands degenerate.

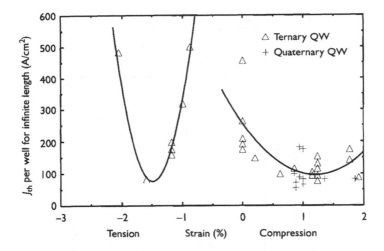

Figure 5.17 Summary of threshold current density, J_{th}, per well deduced for infinite
cavity length 1.5 μm lasers versus the strain in the InGaAs(P) quantum
wells, using data reported in the literature (after Thijs *et al.* © 1994 IEEE).

5.8 Tunnelling structures and devices

When a classical particle is incident on a barrier of height V, it has a 100
per cent probability of transmission if its energy $E > V$, and will be reflected
with 100 per cent probability if $E < V$. By contrast, this is not the case in
quantum mechanics where, because of its wave-like nature, an incident
electron has a finite probability of tunnelling through a thin barrier even
when $E < V$.

The requirements for creating tunnelling structures in III–V semiconductors are not very different to those for quantum wells: a single barrier for
tunnelling can be formed by sandwiching a thin layer of wide band gap
material between two narrower gap regions, as illustrated in fig. 5.18(a).
Because the tunnelling probability, and hence the current, varies exponentially with both barrier height and width, and can also be modified
by impurities, careful growth control is considerably more critical for
tunnelling structures than for quantum well devices, so that tunnelling
devices have not been as widely commercialised. Nevertheless, much elegant physics and many useful effects have been demonstrated in tunnelling
structures, both for electronic and optoelectronic applications (Kelly 1995).

We first review here the principles of quantum mechanical tunnelling
calculations and then conclude this chapter by describing briefly two
of the most significant tunnelling devices: first, double barrier resonant
tunnelling devices, which can display negative differential resistance

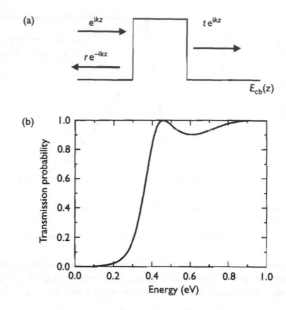

Figure 5.18 (a) Variation of the conduction band edge energy through a single barrier tunnelling structure. A plane wave e^{ikz} incident from the left has a finite probability of being reflected from ($r\,e^{-ikz}$) or transmitted ($t\,e^{ikz}$) through the barrier. (b) Variation of the transmission coefficient $t(E)$ as a function of incident electron energy, E, for an electron in GaAs ($m^* = 0.067$) incident on a barrier 60 Å wide and 0.3 eV high (equivalent to $x \sim 0.35$ in an $Al_xGa_{1-x}As$ barrier).

over a wide temperature range, and, second, quantum cascade laser structures, where population inversion is achieved by engineering the electron confined state energies and tunnelling rates in multiple barrier structures.

We consider in fig. 5.18(a) an electron of energy E incident from the left, with wavefunction e^{ikz} on a barrier of height V ($>E$) and thickness d, where we have again defined $k^2 = 2m^*E/\hbar^2$. When reflection and transmission are included at each interface, the wavefunction takes the form

$$\psi(z) = e^{ikz} + re^{-ikz} \qquad z < 0$$
$$= Ae^{\kappa z} + Be^{-\kappa z} \qquad 0 < z < d$$
$$= te^{ikz} \qquad z > d \qquad (5.41)$$

where $r\,e^{-ikz}$ then describes the wave reflected to the left, $t\,e^{ikz}$ is the transmitted wave, and $Ae^{\kappa z} + Be^{-\kappa z}$ describes the wave in the barrier region, with $\kappa^2 = 2m^*(V-E)/\hbar^2$, and assuming the same effective mass m^* in

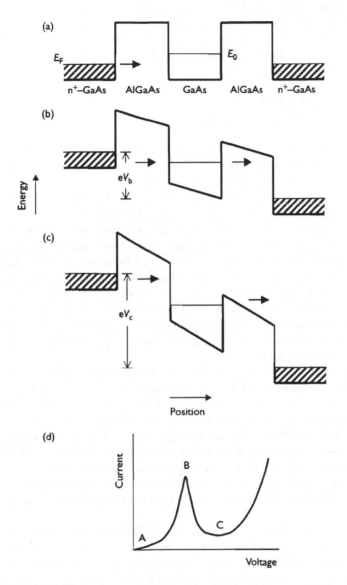

Figure 5.19 (a)–(c) The variation in the conduction band line-up with increasing applied bias in a double barrier structure. (d) The current–voltage characteristic for such a structure. The current is peaked about the point B where resonant tunnelling is possible, and can show a strong negative differential resistance as the bound state moves off-resonance in (c).

each region. We have four unknown coefficients in eq. (5.41), which we can determine by requiring the wavefunction, ψ, and its derivative, $d\psi/dz$ to be continuous at $z = 0$ and $z = d$, the two interfaces. The most important quantity for us is the transmission coefficient, $t(E)$, which, it can be shown (e.g. Schiff 1968), is given by

$$t(E) = \frac{\exp(-ikd)(1 - \phi^2)}{\exp(\kappa d) - \phi^2 \exp(-\kappa d)} \tag{5.42}$$

with $\phi = (\kappa + ik)/(\kappa - ik)$. This is dominated for $E < V$ by the $e^{\kappa d}$ term in the denominator, so that the transmission amplitude then decreases with increasing $V - E$, as illustrated in fig. 5.18(b).

We now consider a double barrier structure, where the two barriers are separated by a thin layer of narrow band gap material, as illustrated in fig. 5.19(a). If the barriers are wide enough, we can clearly view the central region as a quantum well, with a set of confined state energies at E_1, E_2 etc. Even with narrow barriers, there will be resonant states in the well region. We can calculate the tunnelling probability through the double barrier structure as a function of incident energy, E. The tunnelling rate again decreases exponentially with increasing $V - E$, except that near the confined state energies, E_1 etc. resonant tunnelling occurs. For a symmetric structure such as in fig. 5.19(a), the tunnelling probability equals 1 at the confined state energies.

Figure 5.19 illustrates how this effect can be used to achieve negative differential resistance (NDR). The double barrier structure is designed with n-doped layers on either side of the undoped double barrier region. For zero applied bias, electrons incident, for example, from the left are off-resonance (fig. 5.19(a)). When a voltage is applied, much of the applied bias is dropped across the undoped barrier region, so that resonant tunnelling becomes possible, leading to a sharp increase in current (fig. 5.19(b)). With further applied bias, the incident electrons move off-resonance (fig. 5.19(c)), giving a NDR region, where the current decreases for increasing applied voltage.

The current–voltage characteristics of NDR devices are unstable, switching between high and low current values. As the resonant tunnelling process is extremely fast, very high speed oscillation is possible in electronic devices: resonant tunnelling diodes have been demonstrated operating at frequencies of over 700 GHz (Brown *et al.* 1991).

Turning now to optical devices, it was first realised in the early 1970s that it should be possible to achieve lasing by exciting a population inversion between two confined sub-bands in a quantum well structure (Kazarinov and Suris 1971). The chief difficulty is to maintain the population inversion. This problem can be overcome through the clever design of tunnelling structures. The wavelength in such a structure is then determined by

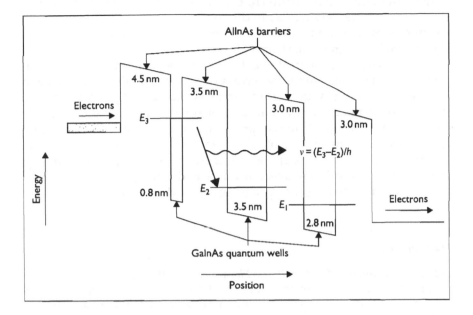

Figure 5.20 Conduction band edge profile of the key section of the first quantum cascade laser to be demonstrated. The structure is designed so that electrons can tunnel efficiently from the left into level 3, leading to a build-up of carriers in this level. Tunnelling between 3 and 2 is slow. Electrons also tunnel efficiently out of levels 2 and 1, leaving them effectively empty and resulting in a population inversion between levels 3 and 2.

the spacing between the confined energy levels, which can be controlled through the layer thicknesses. A tunnelling-based laser was first demonstrated by Federico Capasso and co-workers at Bell Laboratories in New Jersey in 1994 (Faist *et al.* 1994). Laser emission has since been achieved over a wide wavelength range (at least between 4 and 11 μm), including room temperature operation.

Figure 5.20 illustrates the conduction band edge profile of their original device when under forward bias. It consisted of three GaInAs quantum wells sandwiched between four AlInAs barriers, all grown lattice-matched on an InP substrate. There was only one confined sub-band in each well. Under forward bias, electrons tunnel into the first quantum well, leading to a build-up of charge in the highest of the energy levels, E_3. Tunnelling between 3 and 2 is designed to be slow. Electrons tunnel efficiently out of levels 2 and 1, leaving them effectively empty and giving a population inversion between levels 3 and 2. Stimulated radiative recombination between these levels gives laser emission at a frequency $v = (E_3 - E_2)/h$. Because the tunnelling rate from level 3 to 2 is still high in comparison with the radiative recombination rate, each device must contain many such structures in series, giving rise to the name quantum cascade laser, with electrons that tunnel out of level 1 in one structure then cascading on to tunnel into level 3 in the next structure.

The initial device contained twenty five such sections in series, the complete laser containing some 500 layers of precisely defined composition and width. The quantum cascade laser is undoubtedly a *tour de force* in the understanding and application of semiconductor band structure engineering, demonstrating the growth control, physical understanding and engineering possibilities of advanced low-dimensional semiconductor structures. It also opens up the possibility of extending semiconductor lasers across a broad new spectral range and into applications such as miniature radar, remote sensing, and laser-based pollution monitoring.

References

Adams, A. R. and E. P. O'Reilly (1992) *Physics World* **5(10)** 43.

Adams, A. R., E. P. O'Reilly and M. Silver (1998) 'Strained layer quantum well lasers', p. 123 in *Semiconductor Lasers I, Fundamentals*, ed. E. Kapon, Academic Press, New York.

Bastard, G. (1988) *Wave Mechanics Applied to Semiconductor Heterostructures*, Les Editions de Physique, Courtaboeuf.

Bernard, M. G. A. and B. Duraffourg (1961) *Phys. Stat. Sol.* **1** 699.

Biasiol, G. and E. Kapon (1998) *Phys. Rev. Lett.* **81** 2962.

Brown, E. R., J. R. Soderstrom, R. D. Parker, L. J. Mahoney, K. M. Molvar and T. C. McGill (1991) *Appl. Physics Letters* **58** 2291.

Cage, M. E., R. F. Dziuba and B. F. Field (1985) *IEEE Trans. Instrum. Meas.* **34** 301.

Eberl, K. (1997) *Physics World* **10(9)** 47.

Faist, J., F. Capasso, D. L. Sivco, C. Sirtori, A. L. Hutchinson and A. Y. Cho (1994) *Science* **264** 553.

Fang, F. F. and W. E. Howard (1966) *Phys. Rev. Lett.* **16** 797.

Hook, J. R. and H. E. Hall (1991) *Solid State Physics*, 2nd edn, Wiley, Chichester.

Kazarinov, R. F. and R. A. Suris (1917) *Sov. Phys. Semicond.* **5** 207.

Kelly, M. J. (1995) *Low-dimensional Semiconductors: Materials, Physics, Technology, Devices*, Clarendon Press, Oxford.

Kivelson, S., D.-H. Lee and S.-C. Zhang (March 1996) *Scientific American* p. 64.

Lambert, B. *et al.* (1987) *Semiconductor Science and Technology* **2** 491.

O'Reilly, E. P. (1989) *Semiconductor Science and Technology* **4** 121.

Schiff, L. I. (1968) *Quantum Mechanics*, 3rd edn, McGraw-Hill, Tokyo.

Stormer, H. L., A. C. Gossard and W. Wiegmann (1982) *Solid State Commun.* **41** 707.

Thijs, PJ. A., L. F. Tiemeijer, J. J. M. Binsma and T. van Dongen (1994) *IEEE J. Quantum Electronics* **30** 477.

von Klitzing, K. (1986) *Rev. Mod. Phys.* **58** 519.

Willett, R., J. P. Eisenstein, H. L. Stormer, D. C. Tsui, J. C. M. Hwang and A. C. Gossard (1987) *Phys. Rev. Lett.* **59** 1776.

Problems

5.1 Estimate the maximum width for a GaAs quantum well if it is desired that the highest HH confined state is separated by at least 40 meV from the highest LH confined state. You may assume that the barrier due to the neighbouring confining layers is effectively infinite and that the relative effective mass of GaAs heavy holes and light holes is given by $m^*_{HH} = 0.35$ and $m^*_{LH} = 0.082$, respectively. It may also be useful to use $\hbar^2/m = 7.62$ eV Å^2.

5.2 The direct band gap of Al$_x$Ga$_{1-x}$As varies with composition x approximately as (Lambert 1987)

$$E_g(x) = (1.52 + 1.34x)\text{eV}$$

and the ratio of the conduction band to valence band offset, $\Delta E_c : \Delta E_v$, has been determined to be about 65 : 35. By rewriting eq. (5.3a) as

$$L = (2/k)\tan^{-1}(\kappa/k)$$

calculate the three GaAs well widths L_e, L_l, and L_h for which the lowest confined electron, LH, and HH states, respectively, have a confinement energy of 25 meV when $x = 0.2$ in the barrier. (Assume $m^*_e = 0.067, m^*_{HH} = 0.35$, and $m^*_{LH} = 0.082$ in both the well and barrier layers.)

5.3 It can be shown (e.g. Schiff 1968) that the energy levels of the (spherically symmetric) s states in a spherical quantum dot of radius a can be found by solving

$$-k\cot(ka) = \kappa$$

where $E = \hbar^2k^2/2m^*$, $(\Delta E_c - E) = \hbar^2\kappa^2/2m^*$, ΔE_c is the band offset and m^* the carrier effective mass in the dot and barrier. Show that there will be no bound state in a spherical quantum dot unless $\Delta E_c a^2 > \pi^2\hbar^2/8m^*$, and that there is only one bound s state if $\pi^2\hbar^2/8m^* \leq \Delta E_c a^2 < 9\pi^2\hbar^2/8m^*$.

5.4 Estimate the minimum radius required of an InAs spherical quantum dot embedded in a GaAs matrix in order to ensure there will be one bound electron state in the quantum dot. Estimate also the radius at which a second state will become bound. You may assume that the conduction band offset between the InAs dot and GaAs is approximately $\Delta E_c = 0.5$ eV, and that the electron effective mass $m^* = 0.04$ in the dot and barrier layers. (A more accurate calculation would take account of the different electron effective masses in the dot and barrier, but would not significantly change the estimated radii.)

5.5   Because the Hall effect involves two voltage and current components, Ohm's Law needs to be generalised to a tensor form when considering the Hall effect:

$$\begin{pmatrix} V_x \\ V_y \end{pmatrix} = \begin{pmatrix} R_{xx} & R_{xy} \\ R_{xy} & R_{yy} \end{pmatrix} \begin{pmatrix} I_x \\ I_y \end{pmatrix}$$

By calculating the inverse of the above resistance matrix, show that the elements of the conductance matrix, describing how current depends on voltage, are given by

$$G_{xx} = \frac{R_{yy}}{R_{xx}R_{yy} - R_{xy}^2}; \qquad G_{xy} = \frac{-R_{xy}}{R_{xx}R_{yy} - R_{xy}^2}$$

From symmetry, we expect $R_{xx} = R_{yy}$, leading to the apparent paradox that the longitudinal conductance, $G_{xx} = 0$, at the same time as the longitudinal resistance, $R_{xx} = 0$.

5.6   Consider a two-dimensional structure for which the conduction (valence) band density of states per unit area, $g_{c(v)}(E) = m^*_{c(v)}/\pi\hbar^2$, with the band edges at $E = E_c$ and $E = E_v$, respectively. Use eqs (5.37) and (5.38) for the quasi-Fermi levels to show that the conduction electron density, n, depends on the quasi-Fermi level energy F_c as

$$n = \frac{m^*_c kT}{\pi\hbar^2} \ln[e^{(F_c - E_c)/kT} + 1]$$

with an equivalent expression for the valence hole density, p. Invert this expression to show that the Fermi energy varies with carrier density n as

$$F_c - E_c = kT \ln\left[\exp\left(\frac{n\pi\hbar^2}{m^*_c kT} \right) - 1 \right]$$

5.7   Consider an ideal quantum well laser for which the conduction (valence) band density of states per unit area, $g_{c(v)}(E) = m^*_{c(v)}/\pi\hbar^2$, with the band edges at $E = E_c$ and $E = E_v$, respectively. It can be shown that the peak gain, g_{max}, occurs at the band edge energy, and is given by

$$g_{max} = G_0(f_c - f_v)$$

where f_c and f_v are the values of the quasi-Fermi functions at the conduction and valence band edges respectively and G_0 is a constant.

Show that if charge neutrality is maintained in the laser active region ($n = p$) then the peak gain depends on carrier density n as

$$g_{max} = G_0 \left[1 - \exp\left(-\frac{\pi \hbar^2 n}{m_v^* kT} \right) - \exp\left(-\frac{\pi \hbar^2 n}{m_c^* kT} \right) \right]$$

Show that if $m_c^* = m_v^*$, then the transparency carrier density, n_0, (for which $g_{max} = 0$) is given by

$$n_0 = \frac{m_c^* kT}{\pi \hbar^2} \ln 2$$

This shows that the transparency (and hence threshold, n_{th}) carrier density increases with the band edge effective mass in a quantum well laser, and also that n_0 increases approximately linearly with temperature.

   Justify why the transparency carrier density will vary as $n_0 \sim (k_B T)^{D/2}$ in a D-dimensional structure, and hence why the threshold current density is predicted to have a lower temperature dependence as the dimensionality D is reduced in the active region of an ideal semiconductor laser.

Chapter 6

Diamagnetism and paramagnetism

6.1 Introduction

When a magnetic field is applied to an isolated atom, the electrons and nucleus will respond to the applied field, leading to an induced magnetisation. An isolated atom generally has only a weak response to the applied field, with the atom displaying either diamagnetic and/or paramagnetic behaviour, depending on the electron configuration. By contrast, solids show a much wider variety of magnetic responses, including ferromagnetism, antiferromagnetism, and ferrimagnetism. These last three arise due to interactions between electrons on different atoms within the solid.

The aim of this chapter and the next one is, first, to provide an overview of the causes of the different types of magnetic behaviour and, second, to consider some of their consequences. We begin here by considering the behaviour of paramagnetic and diamagnetic atoms and solids, where the response of each atom or ion can generally be viewed as independent of all the others. Most solids are paramagnetic or diamagnetic, with a very weak response to an applied magnetic field. The next chapter considers the more interesting case of ferro-, ferri-, and antiferromagnets, where the additional interactions occurring between electrons on neighbouring ions enable the electrons to respond in a cooperative manner to an applied field, allowing much stronger magnetic effects.

6.2 Magnetisation

The magnetisation, M, of a material can be defined as its magnetic dipole moment per unit volume

$$M = Nm \qquad (6.1)$$

where m is the average magnetic dipole moment per atom or molecule, and N is the number of atoms or molecules per unit volume. M is the magnetic analogue of the electric polarisation P, where P is defined as the electric dipole moment per unit volume. Just as the polarisation response of many materials to an electric field E can be described by the electric susceptibility, χ_e, with $P = \varepsilon_0 \chi_e E$, so too we define the magnetic susceptibility χ_m relating the magnetisation M to the magnetic field intensity H as

$$M = \chi_m H \tag{6.2}$$

The steady-state magnetic field intensity H can be viewed as the field due to an externally applied current, with

$$\nabla \times H = J_f \tag{6.3}$$

where J_f is the external current density. The force F on a charge q moving with velocity v is then given by

$$F = q(v \times B) \tag{6.4}$$

where B is a directly measurable quantity, and is formally defined as the magnetic induction, or magnetic flux density. In practice, few physicists ever refer to the 'magnetic induction', preferring to talk instead about 'the magnetic field, B'. We shall follow this general trend here. The magnetic field, B depends on M and H as

$$B = \mu_0(H + M) \tag{6.5}$$

where μ_0 is the permeability of free space. The definitions of electric susceptibility, χ_e, and magnetic susceptibility, χ_m, are not exactly equivalent. The electric susceptibility is defined as the constant of proportionality linking two directly measurable quantities, the polarisation P and the electric field E. By direct analogy, we might then expect the magnetic susceptibility to have been defined as χ_B, with

$$\mu_0 M = \chi_B B \tag{6.6}$$

and the susecptibility then linking the two measurable quantities, M and B. It is to a certain extent easier to calculate χ_B than χ_m for both paramagnetic and diamagnetic materials, which we do later in this chapter. It can be shown by combining eqs (6.2), (6.5), and (6.6) that the two definitions of magnetic susceptibility are in any case directly related to each other by

$$\chi_m = \frac{\chi_B}{1 - \chi_B} \tag{6.7}$$

so that for the small values of χ_B found in paramagnetic and diamagnetic solids it is in any case a reasonable approximation that $\chi_m = \chi_B$.

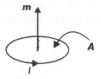

Figure 6.1 An electron moving in a classical orbit of area **A** has magnetic moment $m = IA$ due to its orbital motion, where I is the current flowing, and the moment points perpendicular to the loop.

6.3   Magnetic moment of the electron

Consider an electron with charge e moving in a classical circular orbit of area A (fig. 6.1). It can be shown that the electron has a magnetic moment m due to its motion, given by

$$m = IA \qquad (6.8)$$

where I is the current flowing in the closed loop, and the vector A has magnitude equal to the loop area and points perpendicular to the loop, with the current flowing clockwise when looking along the direction of A. The current I equals the charge passing any point on the loop per unit time, and is given by

$$I = -e\,\frac{\omega}{2\pi} \qquad (6.9)$$

where ω is the angular frequency of the electron motion, and the minus sign follows from the negative charge of the electron. Substituting in eq. (6.8), the magnitude of the electron's magnetic moment due to its orbital motion is then given by

$$m = -\left(e\frac{\omega}{2\pi}\right) \cdot (\pi r^2)$$

$$= \frac{-e}{2m_e} \cdot (m_e r^2 \omega) \qquad (6.10)$$

where m_e equals the electron mass and $Q = m_e r^2 \omega$ is its angular momentum. This result remains true in quantum mechanics, where the angular momentum is quantised, as shown in Appendix B, with orbital quantum number l. If we consider an externally applied field, B, directed along the z-direction, then the component of angular momentum Q_z along the field direction is given by

$$Q_z = \hbar l_z \qquad (6.11)$$

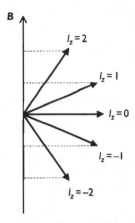

Figure 6.2 The total angular momentum of an electron with orbital quantum number $l = 2$ equals $\hbar\sqrt{l(l+1)} = \sqrt{6}\hbar$. When placed in an applied field **B**, the component of angular momentum along the field direction, l_z, can take the values $-2\hbar$, $-\hbar$, 0, $\hbar$ and $2\hbar$, as illustrated here.

where l_z is an integer, and $-l \le l_z \le l$. This is illustrated for $l = 2$ in fig. 6.2. The energy, E_m, of the magnetic moment in the magnetic field **B** depends on the orientation of the moment with respect to the applied field both classically and quantum mechanically as

$$E_m = -\mathbf{m} \cdot \mathbf{B} \tag{6.12}$$

Substituting eqs (6.10) and (6.11) in eq. (6.12) we then find that the dipole energy is quantised, with

$$E_m = -\mathbf{m} \cdot \mathbf{B} = \frac{e}{2m_e}(\hbar l_z)B$$

$$= \frac{e\hbar}{2m_e}l_z B = \mu_B l_z B \tag{6.13}$$

where $\mu_B = e\hbar/2m_e$, is defined as the Bohr magneton, and can be regarded as the quantised unit of electron magnetic moment.

In addition to orbital angular momentum, an electron also has angular momentum associated with its intrinsic spin, with the spin component, s_z, along the magnetic field direction given by $s_z = \pm\frac{1}{2}$, depending on whether the spin is aligned with or against the applied field. The component of spin magnetic moment, m_{s_z}, along the field direction is given by

$$m_{s_z} = -g_0\mu_B s_z \tag{6.14}$$

where g_0 is referred to as the splitting factor, with $g_0 = 2(1 + e^2/4\pi\varepsilon_0 hc) = 2.0023$, which we can assume equals 2 for further analysis. The inclusion of both c and h in the definition of g_0 indicates that the spin magnetic moment is an effect whose explanation requires a combination of both relativity (c) and quantum mechanics (h).

6.4 Diamagnetism in atoms and solids

The orbital motion of a single electron in an atomic orbital can be regarded as equivalent to that of a classical current loop for which the resistance, $R = 0$. We justify this point further in Chapter 8 when we consider super-conducting materials but for now it is sufficient to note that an electron in a particular orbital of an isolated atom continues its motion *ad infinitum*, as if $R = 0$, until it makes a transition to a completely different state, where its current may be different, but is again constant, with the resistance still equal to zero.

The application of a magnetic field B changes the magnetic flux ϕ through the loop. The total flux is given by

$$\phi = \int_A B \cdot dA \tag{6.15}$$

where the integral A is over the area of the loop. There will, from Lenz's law, be an induced back e.m.f., $\mathcal{E}(t)$, during the time, t, that the field B is changing, given by

$$\mathcal{E}(t) = -\frac{d\phi}{dt} \tag{6.16}$$

This leads to a change in the current flow, and hence in the magnetic moment, giving a (small) induced magnetic field directed opposite to the applied field. As the loop resistance $R = 0$, the change in the loop current will persist until B is removed, giving an induced magnetic moment which points against the applied field B. The diamagnetic susceptibility, χ_{dia} will then be negative. From eq. (6.15) for the magnetic flux, ϕ, we expect that χ_{dia} will depend on the sum of the areas of the occupied orbitals in each atom, with the sum of the areas being dominated by the contribution from the valence electrons, as the core electrons have considerably smaller radii. Calculations outlined in more detail in several other textbooks (e.g. Hook and Hall 1991; Ashcroft and Mermin 1976) show that the diamagnetic susceptibility is independent of temperature and is given by

$$\chi_{dia} = -\mu_0 \frac{Ne^2}{6m_e} \sum_{i=1}^{n_v} \langle R_i^2 \rangle \tag{6.17}$$

where N is the number of atoms per unit volume, n_v is the number of valence electrons per atom, and $\langle R_i^2 \rangle$ is the mean square radius of the ith electron.

We can make an order of magnitude estimate of a typical value for χ_{dia} by setting $\langle R_i^2 \rangle$ equal to the Bohr radius squared, $a_0^2 = (5.3 \times 10^{-10} \, \text{m})^2$, assuming four valence electrons per atom, and $N = 10^{29} \, \text{m}^{-3}$, a typical solid density, which gives $\chi_{dia} = -7 \times 10^{-6}$, a number small compared with one. All solids have a diamagnetic contribution to their magnetic susceptibility, but as this is typically of order -10^{-5} to -10^{-6}, it is a very weak effect overall, certainly in comparison with typical electric susceptibilities. This explains in large part why solids generally interact with electromagnetic waves predominantly through the electric field.

6.5 Langevin (classical) theory of paramagnetism

We saw in Section 6.3 how an individual electron has an associated magnetic moment. For some atoms and most solids, the sum of the magnetic moments associated with all the electrons cancels exactly. The atom or solid then has no net magnetic moment, and the material will be diamagnetic, with a weak response to any applied field.

Consider instead the case where some atoms in a gas or solid have a net magnetic moment, but that the interactions between these atoms are sufficiently weak that they behave essentially independently of each other. In the absence of an applied field, B, the magnetic moments will point at random in all directions, giving no net magnetisation, that is, $M = 0$ for $B = 0$. When a finite field, B, is applied the magnetic moments will tend to align with the magnetic field, to reduce their magnetic energy, E_m. This alignment will, however, be opposed by the thermal agitation of the spins. As the thermal energy increases with increasing temperature, we expect the net induced magnetisation to decrease monotonically with temperature, giving a temperature-dependent paramagnetic susceptibility, $\chi_{par}(T)$. We shall see below that the paramagnetic susceptibility is generally stronger than the diamagnetic susceptibility. Nevertheless, a truly paramagnetic material still has a very weak overall response to an applied field B.

Paramagnetism is displayed by some molecular gases and liquids, including O_2 and NO. It is also observed in transition metal and rare earth salts, such as $MnSO_4(NH_4)_2SO_4 \cdot 6H_2O$, where each transition metal (manganese, Mn) atom has a net magnetic moment, and the magnetic moments are sufficiently separated that they interact only very weakly with each other. In addition, most metals display a temperature-independent paramagnetic susceptibility, as we discuss further in Section 6.8.

We use classical thermodynamics to first estimate the paramagnetic susceptibility of a set of classical spins, which are free to point in any direction.

We then introduce Hund's rules in the next section, which describe how to determine the net magnetic moment of an isolated atom. This will enable us to determine the susceptibility in the quantum mechanical case.

We saw in eq. (6.12) how the magnetisation energy of a magnetic dipole with magnetic moment m aligned at an angle θ to an applied field B is given by

$$E_m = -mB \cos \theta \tag{6.12}$$

implying that the energy is minimised when the magnetic moment is along the field direction. If we assume that there are N dipoles per unit volume, then the number of dipoles per unit volume, dN, aligned between θ and $\theta + d\theta$ (i.e. in the solid angle $d\Omega = 2\pi \sin \theta \, d\theta$ of fig. 6.3) is given using Maxwell–Boltzmann statistics by

$$dN = C_N \exp(-E_m/kT)(2\pi \sin \theta \, d\theta) \tag{6.18}$$

where C_N is a constant of proportionality, which we must evaluate, and $\exp(-E_m/kT)$ is the Boltzmann probability function, describing how the probability of a given state being occupied decreases exponentially with increasing energy. We know that every moment must be aligned at an angle θ between 0 and π with respect to the applied field. We can, therefore, calculate the total number of moments per unit volume, N, by integrating eq. (6.18) between 0 and π:

$$N = \int_{\theta=0}^{\theta=\pi} C_N e^{(mB/kT) \cos \theta} 2\pi \sin \theta \, d\theta \tag{6.19}$$

This integral can be solved by letting $x = mB/kT$ and then making the substitution $y = x \cos \theta$, so that $dy = -x \sin \theta \, d\theta$, and

$$N = \frac{2\pi C_N}{x} \int_{-x}^{x} e^y \, dy = \frac{4\pi C_N}{x} \sinh x \tag{6.20}$$

Re-arranging eq. (6.20), we then calculate that the constant of proportionality, C_N, is given by

$$C_N = \frac{xN}{4\pi \sinh x} \tag{6.21}$$

We note that for small $x (mB \ll kT) \sinh(x) \approx x$, and $C_N \to N/4\pi$, with the N spins then equally distributed over all directions. The net magnetisation M in the field direction can be found as a function of x using

$$M = N\langle m \rangle = \int_0^\pi (m \cos \theta) dN$$

$$\implies M = \frac{xN}{4\pi \sinh x} \int_{\theta=0}^{\theta=\pi} m \cos \theta \, e^{x \cos \theta} 2\pi \sin \theta \, d\theta \tag{6.22}$$

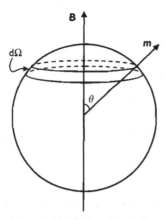

Figure 6.3 The magnetisation energy, E_m, of a magnetic moment m aligned at an angle θ to an applied field **B** is given by $E_m = -mB\cos\theta$. Spins aligned between the angles θ and $\theta + d\theta$ fill a solid angle $d\Omega = 2\pi \sin\theta \, d\theta$.

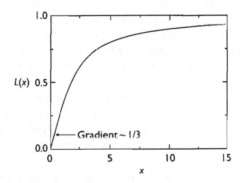

Figure 6.4 Variation with x of the Langevin function, $L(x) = \coth x - 1/x$, which describes the field dependence of the magnetisation in a classical paramagnet. $L(x) \simeq \frac{1}{3}x$ for small x, and goes to 1 at large x.

This can again be solved by letting $y = x\cos\theta$, giving the integral of the function $y\,e^y$, from which we determine the net magnetisation M as

$$M = Nm(\coth x - 1/x) \qquad\qquad (6.23)$$

where $L(x) = \coth x - 1/x$ is known as the Langevin function.

Figure 6.4 shows the variation of the Langevin function with x. $L(x)$ increases linearly for small values of x, when the magnetic energy is much smaller than the thermal energy, that is, $mB \ll kT$. It can be shown that

$L(x) \simeq \frac{1}{3}x$ for $x \ll 1$, so that the net magnetisation, $M = NmL(x)$, is then given by

$$M = Nm\frac{1}{3}\frac{mB}{kT} \tag{6.24}$$

which can be rewritten as

$$\mu_0 M = \frac{N\mu_0 m^2}{3kT}B = \chi_{par}B \tag{6.25}$$

with the paramagnetic susceptibility, χ_{par}, then given by

$$\chi_{par} = \frac{N\mu_0 m^2}{3kT} \tag{6.26}$$

Equation (6.26) predicts that the paramagnetic susceptibility varies inversely with temperature, so that we often write

$$\chi_{par} = \frac{C}{T} \tag{6.27}$$

This is known as the Curie Law, with the material-dependent parameter, C, then called the Curie constant. We can use eq. (6.26) to estimate a typical magnitude for the paramagnetic susceptibility. If we assume $N = 10^{28}\,\mathrm{m}^{-3}$ magnetic moments per unit volume, each with magnitude equal to the Bohr magneton, $m = \mu_B$, we then find for a typical paramagnetic susceptibility that $\chi_{par} \sim 10^{-4}$ at room temperature, again small compared with electric susceptibilities, χ_e.

We can confirm the validity of the linear approximation used in eqs (6.25)–(6.27) by noting that the magnetic energy associated with the Bohr magneton, $\mu_B B$, is only 0.6 meV even for an applied field as large as say $B = 10\,\mathrm{T}$. This is indeed small compared with the thermal energy, as $kT = 25$ meV at room temperature.

The Curie law generally breaks down at very low temperatures, both because the parameter $x = \mu_B B/kT$ becomes large, and also because of the existence of very weak interactions between neighbouring magnetic moments, which can give rise to ferro- or antiferromagnetic effects at the lowest temperatures. Figure 6.4 shows that even in a perfect paramagnetic material, the magnetisation saturates at high magnetic fields. $L(x) \to 1$ when $x \to \infty$, so that the magnetisation is then given by $M = Nm$, with all of the moments aligned with the applied field.

6.6   Magnetic moments in isolated atoms and ions: Hund's rules

We saw in Section 6.3 that there is a magnetic moment associated with each electron in an isolated atom or ion, due to the orbital angular momentum of

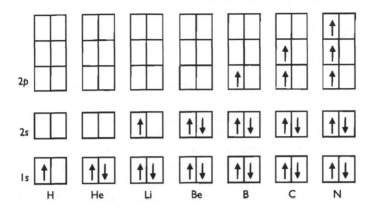

Figure 6.5 Ground state electronic configuration of the first seven atoms in the Periodic Table. The energy is minimised, first, by filling states with lowest principal quantum number, n (labelled '1' and '2' here), then considering lowest available orbital angular momentum, l (labelled 's' and 'p' here), and finally by applying Hund's rules to distribute the electrons among the states of given n and l.

the electron, and its intrinsic spin. The total magnetic moment of the atom or ion is then determined by the way in which the electrons are distributed among the different energy levels. For a given ion, there are often many different ways in which the electrons could in principle be distributed among these levels. The ground state of the atom is then determined by the electron configuration which minimises the total energy.

Figure 6.5 shows the ground state electronic configuration of the first seven atoms in the Periodic Table. It can be seen that the energy levels are filled in a very well defined and systematic manner. The rules describing the ground state electron configuration are referred to as Hund's rules, and have been well established both from analysis of atomic spectra and from advanced calculations. Hund's rules state:

1   The electron spins are arranged so that as many of the electrons as possible have spins which are parallel to each other. This alignment is favoured because of the Pauli Exclusion Principle, one form of which states that there is zero probability of two electrons with the same spin being at the same point at the same time. Hence, electrons with parallel spins tend to have a larger average separation distance than those with anti-parallel spins. This reduces the average electron–electron repulsion energy, and therefore the overall energy. We can define a quantum number, S, to characterise the resultant spin angular momentum of all

the n electrons in the atom, with

$$S = \left| \sum_{i=1}^{n} s_{zi} \right| \tag{6.28}$$

and the total spin angular momentum then being of magnitude $\hbar\sqrt{S(S+1)}$.

2   Subject to the constraint of the first rule, the electrons are then distributed among the possible orbital angular momentum states. Each such state has a quantum number, l_z, associated with its angular momentum component along the quantisation direction. The electrons are distributed among the possible l_z states so that $L = |\sum l_z|$ is maximised, with the resultant total orbital angular momentum then being of magnitude $\hbar\sqrt{L(L+1)}$.

3   The orbital and spin angular momenta are coupled to each other in an isolated ion, and we can define a total quantum number $J = L + S$ associated with this total angular momentum. The magnitude of J is given by

$$J = |L - S| \tag{6.29a}$$

when the shell is less than half-full (in which case L and S point in opposite directions), and

$$J = L + S \tag{6.29b}$$

when the shell is more than half-full (in which case L and S point along the same direction).

Using these rules, the ground state of any isolated ion can be determined, with the z-component of the magnetic moment given by

$$m_z = -g\mu_B J_z \tag{6.30}$$

where g is called the Landé splitting factor, and is given by

$$g = \frac{3}{2} + \frac{S(S+1) - L(L+1)}{2J(J+1)} \tag{6.31}$$

The Landé splitting factor reflects the relative contribution of spin and orbital motion to the total angular momentum. If we have a pure spin state, for which $L = 0$ and $J = S$, then eq. (6.31) gives $g = 2$, as we had earlier in eq. (6.14), while when $S = 0$, and $J = L$, we find $g = 1$, as we had in eq. (6.13) for pure orbital angular momentum.

6.7   Brillouin (quantum mechanical) theory of paramagnetism

The Langevin theory we considered earlier provides a good description of the behaviour of paramagnetic materials. It describes accurately the response to a weak magnetic field, when $x = mB/kT$ is small, and also reproduces the saturation behaviour for large x, where all the magnetic moments tend to align with the applied field, B. However, it tends to underestimate the net magnetisation M for intermediate values of x. The behaviour in this region can be reproduced more accurately using Brillouin theory, which takes account of the quantised nature of the angular momentum and magnetic moment. For an ion whose total angular momentum quantum number is J, the allowed values for the component of magnetic moment along the field direction are given by eq. (6.30), $m_z = -g\mu_B J_z$, where J_z is an integer or half-integer, which takes values $-J, -J+1, \ldots, J-1, J$. We can use a technique similar to the one used before to calculate the net magnetisation M in the field B. We again assume that there are N independent paramagnetic ions per unit volume. The number of ions, $\Delta N(J_z)$, with magnetic moment component $m_z = -g\mu_B J_z$ along the field direction is given by

$$\Delta N(J_z) = C_N e^{m_z B/kT}$$
$$\implies \Delta N(J_z) = C_N e^{-g\mu_B J_z B/kT} \tag{6.32}$$

where C_N is again a constant of proportionality, and $\exp(m_z B/kT)$ is the Boltzmann probability function. We find, summing over the allowed total angular momentum values that

$$N = C_N \sum_{J_z=-J}^{J} e^{-g\mu_B J_z B/kT} \tag{6.33}$$

which can be re-arranged to give

$$C_N = \frac{N}{\sum_{J_z=-J}^{J} e^{-g\mu_B J_z B/kT}} \tag{6.34}$$

The net magnetisation, M, in the field direction can then be found by summing over the allowed energy levels

$$M = N\langle m \rangle = \sum_{J_z=-J}^{J} -g\mu_B J_z \, \Delta N(J_z)$$

so that

$$M = \frac{N}{\sum_{J_z=-J}^{J} e^{-g\mu_B J_z B/kT}} \sum_{J_z=-J}^{J} (-g\mu_B J_z) e^{-g\mu_B J_z B/kT} \qquad (6.35)$$

We leave it as an exercise at the end of the chapter to show that eq. (6.35) can be simplified to give

$$M = Ng\mu_B J B_J(x) \qquad (6.36)$$

where $x = g\mu_B J B/kT$ is again a measure of the relative strength of the magnetic and thermal energy. $B_J(x)$ is referred to as the Brillouin function and takes the form

$$B_J(x) = \frac{2J+1}{2J}\coth\frac{(2J+1)x}{2J} - \frac{1}{2J}\coth\frac{x}{2J} \qquad (6.37)$$

The behaviour of the Brillouin function is qualitatively very similar to the Langevin function. $B_J(x) \to 1$ for large x, so that the magnetisation, M, saturates at $M = Ng\mu_B J$ when the magnetic energy is large compared to the thermal energy. $B_J(x)$ varies linearly with x for small x, where it can be shown that

$$\mu_0 M = \chi_{(J)} B \qquad (6.38)$$

with the paramagnetic susceptibility, $\chi_{(J)}$, given by

$$\chi_{(J)} = \frac{\mu_0 Ng^2 \mu_B^2 J(J+1)}{3kT} \qquad (6.39)$$

This is identical to the classical result for the susceptibility in eq. (6.26), if we identify

$$m = -p\mu_B \qquad (6.40)$$

where $p = g[J(J+1)]^{1/2}$ can be thought of as the effective Bohr magneton number. Figure 6.6 compares the experimentally determined variation of the magnetisation, M, for several paramagnetic salts with the theoretically predicted variation, using the Brillouin function. The measurements were all carried out at low temperature (below 5 K), enabling large values of $x = mB/kT$ to be obtained. It can be seen that excellent agreement was achieved in each case by assuming the net orbital angular momentum, $L = 0$, so that the splitting factor, $g = 2$, with $J = S = 3/2, 5/2$, and $7/2$ respectively.

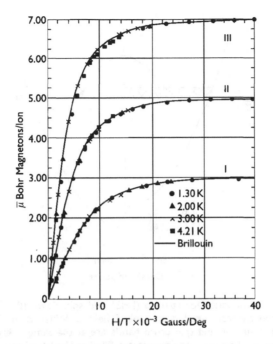

Figure 6.6 Comparison of the experimentally determined variation of the magnetic moment *m* per paramagnetic ion in several salts with the theoretically predicted variation, using the Brillouin function (solid lines) and assuming $L = 0$, with $S = 3/2$ (I, Cr^{3+}), $5/2$ (II, Fe^{3+}) and $7/2$ (III, Gd^{3+}), respectively. (Experimental data after Henry (1952), © 1952 by the American Physical Society.)

6.8 Paramagnetism in metals

So far we have considered only electrons bound to individual atoms. Free electrons in metals also display paramagnetic and diamagnetic behaviour, but their susceptibility is distinguished by being virtually independent of temperature. We saw in Chapter 5 how the density of states for a parabolic band structure with effective mass m_e varies with energy, E, above the band edge as

$$g(E) = 4\pi \left(\frac{2m_e}{h^2}\right)^{3/2} E^{1/2} \tag{6.41}$$

In a metal at zero temperature, all states up to the Fermi energy, E_F, are filled, while the states above E_F are empty. In zero magnetic field, there are equal numbers of electrons with spin up and spin down, implying no net magnetisation. When a field *B* is applied, the electrons with magnetic moment *m* along the field direction are shifted down in energy

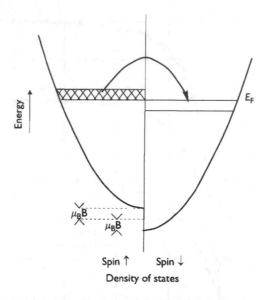

Figure 6.7 Density of states for spin up (LHS) and spin down (RHS) electrons of a free electron metal in an applied field **B**. When **B** = 0, the bottom of the spin up and spin down bands are at the same energy, with equal number of spin up and down states filled to the Fermi energy, E_F. In field **B**, the electrons which occupied the cross-hatched states on the LHS must be transferred into previously empty states on the RHS, giving a net paramagnetic magnetisation.

by $mB = \frac{1}{2}g_0\mu_B B$ while states with m against the field direction shift up by $\frac{1}{2}g_0\mu_B B$, as illustrated in fig. 6.7. The Fermi energy must be the same for both sets of states in thermal equilibrium. Considering the shaded area in fig. 6.7, this occurs by moving $\Delta N = (\frac{1}{2}g_0\mu_B B) \cdot (\frac{1}{2}g(E_F))$ electrons per unit volume from states with moment aligned against the field (above the Fermi energy) to states aligned with moment parallel to the field (below the Fermi energy). The net magnetisation M is then given by

$$M = \frac{1}{2}g_0\mu_B \cdot 2\Delta N = \frac{1}{4}g_0^2\mu_B^2 g(E_F)B \tag{6.42}$$

so that the paramagnetic susceptibility, χ_P, is then given by

$$\chi_P \sim \mu_0\mu_B^2 g(E_F) \tag{6.43}$$

Because thermal energies kT are much smaller than the Fermi energy E_F, the same argument can be applied at finite temperature, and the paramagnetic susceptibility of a metal, referred to as the Pauli spin susceptibility, is then approximately independent of temperature. Because electrons near the Fermi energy are free to move through the metal, they also have a diamagnetic susceptibility associated with this motion. Landau calculated that for a *free-electron* metal (whose density of states is given by eq. (6.41)), the diamagnetic susceptibility, χ_L, is given by $\chi_L = -\frac{1}{3}\chi_P$, so that the net susceptibility for free electrons is then positive and is equal to $\frac{2}{3}\chi_P$. Band structure and related effects modify this result, but nevertheless it is still found that the susceptibility is temperature independent and is, for many metals, of comparable magnitude to that predicted by the Pauli model.

6.9 Floating frogs

It is fascinating to see objects floating without material support or suspension. This became a familiar sight in the 1980s when pellets of the then new high-temperature superconductors were levitated above permanent magnets, and vice versa. We have emphasised throughout this chapter how the magnetic response of most materials is very weak. It is, therefore, surprising to find that ordinary diamagnetic objects can also be levitated in achievable magnetic fields. Figure 6.8 illustrates a frog floating above a 16 T magnet.

Figure 6.8 A small frog floating above a high field superconducting magnet, due to diamagnetic repulsion. (Photograph courtesy of A. Geim, University of Manchester.)

Whether an object will float or not is determined by the balance between the magnetic force and gravity. For a diamagnetic material, the induced magnetic moment, $m_{dia} = MV$, where M is the magnetisation (given by eq. (6.6)) and V the volume, with m_{dia} then given by

$$m_{dia} = \frac{\chi VB}{\mu_0} \tag{6.44}$$

By integrating the work $-dm_{dia} \cdot B$ as the field is increased from zero to B we can obtain the total magnetic energy of the object. Adding this to the gravitational energy, mgz, where m is the mass and z the vertical position coordinate, the total energy E is given by

$$E = mgz - \frac{\chi V}{2\mu_0} B^2 \tag{6.45}$$

For the object to float, the total vertical force, $F_z = -\partial E/\partial z$, must vanish so that

$$-mg + \frac{V\chi}{\mu_0} B \frac{\partial B}{\partial z} = 0 \tag{6.46}$$

The equilibrium condition then becomes

$$B \frac{\partial B}{\partial z} = \frac{\mu_0 \rho g}{\chi} \tag{6.47}$$

which we note involves only the density, ρ, of the levitated object, not its mass. If we take $\rho \sim 10^3 \, \text{kg/m}^3$ and $\chi \sim -10^{-5}$ for a diamagnet or $\chi \sim 10^{-3}$ for a paramagnet, magnetic levitation then requires $B \, \partial B/\partial z \sim 1000 \, \text{T}^2 \, \text{m}^{-1}$ or $10 \, \text{T}^2 \, \text{m}^{-1}$ respectively. Taking $l \sim 0.1 \, \text{m}$ as the typical size a of high field magnet and assuming $\partial B/\partial z \sim B/l$, we find that fields of order 1 or 10 T are sufficient to cause levitation of para- and diamagnets.

We have not addressed here the equilibrium against horizontal displacement – this is more complex, but has been treated by Berry and Geim (1997), who prove that a diamagnet can float in stable equilibrium above a magnet, whereas a paramagnet (which floats beneath the magnet) is always unstable to horizontal movements, accelerating away when displaced from the vertical symmetry axis, as outlined in problems 6.6–6.8 below.

We conclude that magnetic forces are nevertheless generally small. We think of them as significant because they can be comparable to or even exceed gravitational forces. Yet if we assume that there are $N \sim 10^{29} \, \text{atoms/m}^3$ in a diamagnetic solid then from eq. (6.45) the energy stored in a 10 T field is less than 1 μeV per atom – orders of magnitude smaller than the electrostatic interactions we shall be considering in the next chapter.

References

Ashcroft, N. W. and N. D. Mermin (1976) *Solid State Physics*, Holt, Rinehart and Winston, New York.

Berry, M. V. and A. K. Geim (1997) *Eur. J. Phys.* **18** 307.

Henry, W. E. (1952) *Phys. Rev.* **88** 559.

Hook, J. R. and H. E. Hall (1991) *Solid State Physics*, 2nd edn, Wiley, Chichester.

Problems

6.1 Show that the Brillouin function for an ion with total angular momentum $J = \frac{1}{2}$ is of the form

$$\langle \mu \rangle = \tfrac{1}{2} g \mu_B \tanh(x)$$

where $\langle \mu \rangle$ is the mean magnetic moment and $x = g\mu_B J B/kT$. Calculate the paramagnetic susceptibility of a solid containing N such ions per unit volume.

6.2 Establish the form of the Brillouin function for an ion whose total angular momentum equals J:

$$B_J(x) = \frac{2J+1}{2J} \coth \frac{(2J+1)x}{2J} - \frac{1}{2J} \coth \frac{x}{2J}$$

Show that this approaches the classical Langevin function as $J \to \infty$. Show also that the low field susceptibility is equivalent for all values of J to the classical result if we associate the magnetic moment $m = -p\mu_B$ defined in eq. (6.40) with each ion.

6.3 Use eq. (6.17) to calculate the diamagnetic susceptibility of a gas consisting of N hydrogen atoms per unit volume, given that the ground state wavefunction of a hydrogen atom is $\psi(r) = (\pi a_0^3)^{1/2} \exp(-r/a_0)$ where a_0 is the Bohr radius ($a_0 = 0.529$ Å). Given that $J = S = \frac{1}{2}$ for hydrogen, calculate the paramagnetic susceptibility of the gas at room temperature and determine the temperature T for which the net susceptibility is zero.

6.4 Establish that for Cr^{3+} with the configuration $3d^3$ in the unfilled 3d shell, the experimental value 3.8 for p fits the expression $2\sqrt{S(S+1)}$ better than the expression $g\sqrt{J(J+1)}$. This and related experimental data (e.g. in fig. 6.6) provides direct experimental evidence that the orbital angular momentum is generally 'quenched' in paramagnetic transition metal ions, with the magnetic moment then being due just to spin angular momentum.

6.5 The rare earth element Dy, atomic number 66, has in addition to the xenon configuration of filled electron states, two 6s states ($6s^2$) and ten 4f ($4f^{10}$) states filled. It loses the two 6s and one 4f electrons to form the Dy^{3+} ion. Show that the basic level of Dy^{3+} is $^6H_{15/2}$ and verify that the experimentally observed value for p of 10.6 is close to the theoretically expected value.

(Note $L = 3$ for an f shell, which can hold 14 electrons. The notation ^nH_m indicates $n = 2S + 1$, where S is total spin angular momentum; the total orbital angular momentum is given by the capital letter, with $S, P, D, F, G, H, \ldots \equiv 0, 1, 2, 3, 4, 5 \ldots$; and the total angular momentum $J = m$.)

6.6   We stated in Section 6.9 that it is impossible for a paramagnet to float stably in a magnetic field. A necessary condition for stability at point P is that the force F is always directed back towards P, so that $\int_A F \cdot dA < 0$, where A is a surface surrounding P. Hence $\nabla \cdot F < 0$.

a   Show from eqs (6.45) and (6.46) that $\nabla \cdot F < 0$ requires $\chi \nabla^2 B^2 < 0$.

b   Show, using Maxwell's steady-state equations, $\nabla \cdot B = 0$ and $\nabla \times B = 0$, that $\nabla^2 B_x = \nabla^2 B_y = \nabla^2 B_z = 0$ in a steady magnetic field.

c   Hence show that $\nabla^2 B^2 = 2[|\nabla B_x|^2 + |\nabla B_y|^2 + |\nabla B_z|^2] \geq 0$, so that a paramagnet will never float stably in a magnetic field.

6.7   By symmetry, the magnetic field points along the axis at the centre of a circular solenoid, $B = B(0, 0, z)k$, where k is the unit vector along the z-direction. Use Maxwell's equations and Taylor's theorem to show that $B(x, 0, z)$ is given at small x by

$$B(x, 0, z) = -\tfrac{1}{2} x B^{(1)}(0, 0, z)i + (B(0, 0, z) - \tfrac{1}{4} x^2 B^{(2)}(0, 0, z))k$$

where $B^{(n)} = \partial^n B(0, 0, z)/\partial z^n$.

6.8   The sufficient condition for stability of a diamagnet at a point where $F = 0$ along the axis of a solenoid is that $\partial^2 E/\partial x^2 > 0$ (horizontal stability) and $\partial^2 E/\partial z^2 > 0$ (vertical stability). The magnetic field along the axis of a circular current loop varies with distance along the axis as $B(0, 0, z) = B_0/[1 + (z/a)^2]^{3/2}$, where a is the radius of the loop. Show that stable equilibrium can be achieved if B_0 is such that the gravitational and magnetic forces are balanced in the region bounded by $a/\sqrt{7} < z < (2/5)^{1/2}a$.

Chapter 7

Ferromagnetism and magnetic order

7.1 Introduction

Many materials possess an ordered array of magnetic moments at room temperature and above, due to interactions between electrons on neighbouring sites. In a ferromagnet, the moments align in the same direction, as illustrated in fig. 7.1(a), so that the material has a net magnetisation, M. In an antiferromagnet, we again get ordering but the net magnetisation is zero, because half the moments (on one sublattice) align along one direction, while the other half align along the opposite direction (fig. 7.1(b)). Finally, in a ferrimagnet, we have oppositely directed moments on two sublattices, but the moments do not cancel, so giving a net magnetisation (fig. 7.1(c)). These different types of materials, in which cooperative magnetic effects occur, display a very wide range of interesting and useful properties: some respond very rapidly to an applied magnetic field, and are therefore useful for high-power and ultra-fast transformer applications; others display strong hysteresis effects, and are therefore suited as stable permanent magnets or for high density magnetic data storage. We first review here the general model of magnetic ordering, before turning

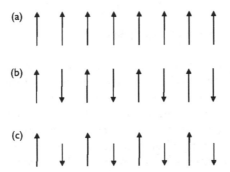

Figure 7.1 Local ordering of magnetic moments in (a) a ferromagnet, (b) an antiferromagnet, and (c) a ferrimagnet.

to consider some of the features which enable the diversity of properties. Finally, we consider some specific applications.

7.2 The exchange interaction

Magnetic ordering in a ferromagnet is *not* due to magnetic interactions between neighbouring dipoles. Consider two neighbouring atoms, each with a magnetic moment equal to the Bohr magneton, μ_B. The magnitude of the magnetic field at one atom due to the other is given approximately by

$$B \sim \frac{\mu_0 \mu_B}{4\pi r^3} \tag{7.1}$$

where r is the separation between the two atoms, $\sim 3\,\text{Å}$ for nearest neighbours. The interaction energy ΔE is then of order $\mu_B B$, and substituting for μ_0 and μ_B in eq. (7.1) we can estimate the direct magnetic interaction between neighbouring atoms as $\Delta E \sim 10^{-6}$ eV, which is considerably less than thermal energies. Direct magnetic interactions are, therefore, too weak to overcome thermal disordering effects, and we must seek an alternative explanation for magnetic ordering. This is provided by the *exchange interaction*.

The exchange interaction is a quantum mechanical effect – it has no classical analogue – and arises due to the electrostatic interaction between electrons, as discussed below. We have already encountered the Pauli Exclusion Principle, which states that no two electrons can occupy the same energy state. An alternative expression of the exclusion principle states that the wavefunction ψ describing two electrons with coordinates r_1 and r_2 and spins s_1 and s_2 must be anti-symmetric when all the coordinates of the two electrons are exchanged, including their position and spin:

$$\psi(r_1, s_1; r_2, s_2) = -\psi(r_2, s_2; r_1, s_1) \tag{7.2}$$

This immediately requires that $\psi = 0$ when $r_1 = r_2$ and $s_1 = s_2$, so that there is zero probability of finding two electrons of the same spin at the same point in space. By contrast, electrons with opposite spin *can* be at the same point. We therefore expect that for two electrons on the same atom, their average separation $\langle |r_1 - r_2| \rangle$ will be larger for parallel spins ($s_1 = s_2$) than for anti-parallel spins ($s_1 = -s_2$). Hence the inter-electron Coulomb repulsion energy $\langle e^2/(4\pi\varepsilon_0|r_1 - r_2|)\rangle$ is smaller for parallel than for anti-parallel spins. This effect is referred to as the exchange interaction. It immediately explains Hund's first rule in Section 6.6, whereby electrons first occupy all states of the same spin in an isolated atom before they start

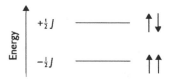

Figure 7.2 Energy level diagram due to the exchange interaction between two electrons on the same atom: the exchange energy, $E_{ex} = -\frac{1}{2}J$ when the electron spins are parallel, while $E_{ex} = +\frac{1}{2}J$ when the spins are anti-parallel.

to occupy states of opposite spin. We can represent the exchange interaction energy, E_{ex}, for two electrons on the same atom by

$$E_{ex} = -2J \, s_1 \cdot s_2$$
$$= -\tfrac{1}{2}J \qquad \text{when } s_{1z} = s_{2z} = \pm\tfrac{1}{2}$$
$$= +\tfrac{1}{2}J \qquad \text{when } s_{1z} = -s_{2z} \qquad\qquad (7.3)$$

as illustrated in fig. 7.2.

The situation is more complicated in molecules and solids, where the exchange interaction also plays a major role in determining the ground state energy. Consider a diatomic molecule with two electrons: the total energy now includes interactions not only between the electrons but also between the electrons and the two nuclei. We saw in Chapter 2 using the independent electron approximation how the ground state wavefunction favours a build-up of charge between the two nuclei. This is best achieved with anti-parallel spins (see fig. 7.3), and indeed explains why most molecules and covalent solids are diamagnetic.

Individual ions can retain a net magnetic moment in some molecules and solids, in particular transition metal ions (e.g. Fe, Co) or rare earth ions (e.g. Nd, Sm, Gd). The direct exchange interaction between electrons on two such ions decreases very rapidly with increasing distance. Indirect exchange interactions are, however, also possible, for example, mediated through the electrons on a shared neighbouring atom or through the spin of conduction electrons in a metal (fig. 7.4). Because competing effects are present, these exchange interactions favour parallel spins and ferromagnetism in some cases, while favouring anti-parallel spins and antiferromagnetism in others.

The exact treatment of the spin interactions in magnetic solids, and their temperature dependence is very complex and difficult, because we are dealing essentially with a many-body effect, which can depend on both local and non-local interactions. Much insight can, however, be obtained from the Heisenberg Hamiltonian, H, which is widely used to describe the

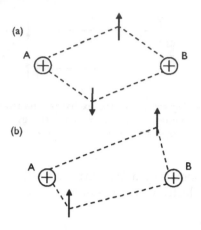

Figure 7.3 (a) When the electron spins are anti-parallel in a diatomic molecule, there can be a high probability of both electrons being between the two nuclei (A and B), maximising the electron–nuclear attractive interaction. (b) By contrast, when the spins are parallel, the electrons tend to avoid each other more, leading to a weaker overall electron–nuclear interaction.

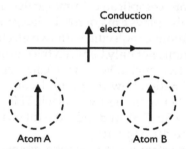

Figure 7.4 Indirect exchange interaction in a metal. The magnetic moment, for example, on atom A has a direct exchange interaction with a free conduction electron, influencing the spin of the conduction electron. The conduction electron then interacts with an electron on atom B, which can enable a large indirect exchange interaction between the magnetic moments on A and B.

spin dependence of the exchange energy of the whole solid:

$$H = -\sum_i \sum_{i \neq j} J_{ij} s_i \cdot s_j \tag{7.4}$$

The contribution to the total exchange energy due to the interaction between atoms i and j, $E_{ex,ij}$, is then given by

$$E_{ex,ij} = -2J_{ij}\, s_i \cdot s_j \tag{7.5}$$

where $\hbar s_i$ and $\hbar s_j$ are the total angular momenta of the electrons on atoms i and j, and the factor of two arises from the double summation in eq. (7.4). We take $J_{ij} > 0$ when parallel spins are favoured, and $J_{ij} < 0$ when an antiparallel alignment is preferred. The magnitude of J_{ij} generally decreases very rapidly with increasing distance, so that it is common to assume $J_{ij} = J$ for nearest-neighbour atoms, and is equal to zero otherwise.

7.3 Ferromagnetism and the Curie temperature

We can, from eq. (7.4), define the exchange energy associated with a particular site i, $E_{ex,i}$, as

$$E_{ex,i} = -\sum_j J_{ij}\, s_i \cdot s_j = -s_i \cdot \sum_j J_{ij} s_j \tag{7.6}$$

The magnetic moment, m_i at site i depends on the total electron angular momentum s_i as $m_i = -g\mu_B s_i$ where g is the Landé g-factor and μ_B the Bohr magneton (eq. 6.30). We can, therefore, rewrite eq. (7.6) as

$$E_{ex,i} = -\frac{m_i}{g\mu_B} \cdot \sum_j \frac{J_{ij} m_j}{g\mu_B} = -\mu_0 m_i \cdot H_{int} \tag{7.7}$$

where

$$H_{int} = \frac{\sum_j J_{ij} m_j}{\mu_0 (g\mu_B)^2} \tag{7.8}$$

has the dimensions of magnetic field intensity, and is referred to as the 'internal field'. H_{int} varies from site to site, because of fluctuations in the local moments. We see, however, that the average internal field is proportional to $\langle m_j \rangle$, the average magnetic moment per ion. As the magnetisation, $M = N\langle m_j \rangle$, where N is the number of magnetic moments per unit volume, we can define the mean internal field, $\langle H_{int} \rangle$ as linearly dependent on M:

$$\langle H_{int} \rangle = \lambda M \tag{7.9}$$

where λ is the constant of proportionality linking $\langle H_{int} \rangle$ and M.

Using this mean field approximation, we then estimate the effects of the neighbouring spins on a given site through the so-called internal field, $\langle H_{int} \rangle$.

In an externally applied field H_0, the net effective field F acting on the magnetic ions is

$$F = H_0 + \lambda M \tag{7.10}$$

We saw in the last chapter for a paramagnetic solid that the magnetisation M depends on the locally applied field F as $M = \chi F$, with the susceptibility χ varying with temperature as $\chi = C/T$ when M is small. We, therefore, write for a ferromagnetic solid at high temperature that

$$M = \frac{C}{T}(H_0 + \lambda M) \tag{7.11}$$

which can be rearranged to give

$$M = \frac{C}{T - T_c}H_0 \tag{7.12}$$

The susceptibility χ is then equal to

$$\chi = \frac{C}{T - T_c} \tag{7.13}$$

where $T_c = \lambda C$ is referred to as the Curie temperature.

We see from eq. (7.13) that because of cooperative interactions between neighbouring spins the susceptibility increases more rapidly with decreasing temperature in a ferromagnetic material than in a true paramagnet, where all the spins respond independently of each other. As the temperature approaches T_c from above, the susceptibility diverges ($\chi \to \infty$). At and below T_c eq. (7.13) breaks down and we get ferromagnetic ordering, with the development of a spontaneous, finite magnetisation M even in the absence of an applied field.

7.4 Spontaneous magnetisation

Because we are dealing with a cooperative effect involving feedback, we have two relations between F and M. The first relation (eq. (7.10)) defines the effective field in terms of the magnetisation, with

$$F = \lambda M \qquad \text{when } H_0 = 0 \tag{7.14}$$

We have in addition a second equation relating how the magnetisation M varies with effective field F and temperature T. If we use the Langevin (classical) theory of paramagnetism, the magnetisation M depends on the magnitude of the field as

$$M = Nm_0\left(\coth x - \frac{1}{x}\right) \tag{7.15}$$

where Nm_0 is the maximum possible magnetisation per unit volume (when all the magnetic moments, each of magnitude m_0, are aligned parallel to each other), and

$$x = \frac{\mu_0 m_0 F}{kT} \tag{7.16}$$

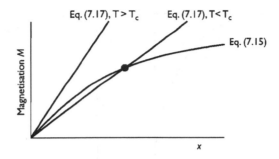

Figure 7.5 The graphical method to determine the spontaneous magnetisation of a ferromagnet, using eqs (7.15) (curve) and eq. (7.17) (straight line). The slope of eq. (7.17) is proportional to the temperature. At high temperatures ($T > T_c$) the straight line and curve only intersect at the origin. For $T < T_c$, the two curves intersect for a finite value of the magnetisation M, and spontaneous magnetisation occurs.

We rewrite eq. (7.14) as

$$M = \frac{kT}{\mu_0 m_0 \lambda} x \tag{7.17}$$

Equations (7.15) and (7.17) are both true so, for a chosen temperature, T, we can plot $M(x)$ in two ways, one from each equation. Any intersections of the two curves indicate values of M which are in fact solutions of both. This procedure can then be repeated for a range of temperatures, resulting in a graph representing the spontaneous magnetisation M as a function of temperature. This is illustrated in fig. 7.5. At high temperatures ($T > T_c$) and with zero external field, the two curves only intersect at $M = 0$. But, once T drops below T_c, the two lines cut both at $M = 0$ and at finite M, giving a spontaneous net magnetisation. (The solution at the origin is unstable; once any spontaneous magnetisation occurs, M will grow to the finite, stable solution.) Figure 7.6 shows the calculated variation of the magnetisation M as a function of temperature below T_c, compared to experimental data for Ni. The overall agreement between the two curves looks excellent, confirming the usefulness of the mean field theory introduced here; there are, however, minor but significant differences between the two curves both near $T = 0$ and near $T = T_c$, to which we will return later.

7.5 Spontaneous magnetisation and susceptibility of an antiferromagnet

The exchange interaction J_{ij} between neighbouring sites is negative in an antiferromagnet, favouring an anti-parallel alignment of neighbouring

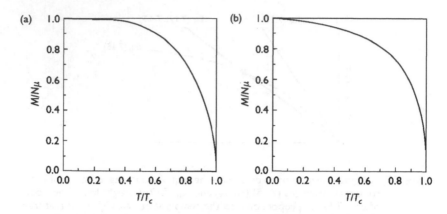

Figure 7.6 (a) The calculated variation relative to its value at $T = 0$ of the spontaneous magnetisation, M, of a ferromagnet for which $J = S = \frac{1}{2}$, compared
to (b) the experimentally determined temperature dependence of the magnetisation in Ni. (From H. P. Myers (1997) *Introductory Solid State Physics*,
2nd edn.)

spins. Complete antiferromagnetic ordering can be obtained at $T = 0$ if
we divide the crystal into two sublattices, as in fig. 7.1(b), with spins pointing up on the first sublattice, which we refer to as sublattice A, and spins
pointing down on the second sublattice (sublattice B). We can define a net
magnetisation M_A for sublattice A and M_B for sublattice B, with the total
magnetisation M then given by $M = M_A + M_B$. We introduce the effective
fields, F_A and F_B, as

$$F_A = H_0 - \lambda M_B \tag{7.18a}$$

$$F_B = H_0 - \lambda M_A \tag{7.18b}$$

describing the influence of sublattice B on A, and vice versa. At high
temperature, we can write

$$M_A = \frac{1}{2}\frac{C}{T}F_A \tag{7.19a}$$

$$M_B = \frac{1}{2}\frac{C}{T}F_B \tag{7.19b}$$

where the factor of $\frac{1}{2}$ is introduced because each sublattice has $\frac{1}{2}N$ magnetic
ions per unit volume. Adding eqs (7.19a) and (7.19b), we find

$$M = \frac{1}{2}\frac{C}{T}(F_A + F_B) \tag{7.20}$$

Substituting eq. (7.18) in (7.20) then gives

$$M = \frac{C}{T}\left(H_0 - \frac{1}{2}\lambda M\right) \tag{7.21}$$

which can be re-arranged as

$$M = \frac{C}{T + \lambda C/2}H_0 \tag{7.22}$$

with the susceptibility χ now equal to

$$\chi = \frac{C}{T + T_N} \tag{7.23}$$

where $T_N = \lambda C/2$ is referred to as the Néel temperature.

The onset of spontaneous magnetisation occurs when we can get finite sublattice magnetisation in zero applied field. Setting $H_0 = 0$ in eqs (7.18) and (7.19), this occurs when

$$M_A = -\frac{1}{2}\frac{C}{T}\lambda M_B \qquad M_B = -\frac{1}{2}\frac{C}{T}\lambda M_A \tag{7.24}$$

Both parts of eq. (7.24) can be satisfied for finite M_A and M_B if $T = \lambda C/2 = T_N$, with mean field theory then predicting spontaneous ordering at the Néel temperature, and for $T < T_N$.

We can calculate the spontaneous magnetisation of the two sublattices below T_N using a similar method to the ferromagnetic case. We apply, for instance, the Langevin (classical) equation for paramagnetism to one of the two sublattices. We write

$$M_A = \frac{N}{2}m_0\left(\coth x_A - \frac{1}{x_A}\right) \tag{7.25}$$

with the second equation linking x_A and M_A derived using the fact that $M_B = -M_A$:

$$x_A = \frac{\mu_0 m_0 F_A}{kT} = \frac{\mu_0 m_0(-\lambda M_B)}{kT} = \frac{\mu_0 m_0 \lambda M_A}{kT} \tag{7.26}$$

Equations (7.25) and (7.26) are equivalent to eqs (7.15) and (7.17) in the ferromagnetic case, implying that the zero-field ordering on each sublattice in an antiferromagnet has a similar temperature dependence to that of a ferromagnet.

Although an antiferromagnet has no net magnetisation ($M = 0$ in zero applied field), antiferromagnetic ordering has been widely studied experimentally, particularly using neutron diffraction. Each neutron has

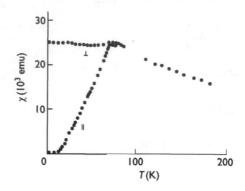

Figure 7.7 The experimentally measured variation of the susceptibility χ of a single domain sample of the antiferromagnetic crystal, MnF_2. The susceptibility, $\chi_{\parallel}$, drops to zero when the applied field is parallel to the magnetisation direction, but remains approximately constant below the Néel temperature when $H \perp M$. (Experimental data after Trapp and Stout, © 1963 by the American Physical Society.)

a magnetic moment, so a spin-polarised neutron beam can detect the additional order in a spontaneously magnetised crystal, with two distinct magnetic sublattices, compared to a disordered crystal where the two sublattices are equivalent.

Because $M = 0$ in an antiferromagnet for zero applied field, it is still possible to define the susceptibility, χ, below the Néel temperature, T_N. The susceptibility is found to depend on the direction of the applied field H_0 relative to the magnetisation direction, as illustrated in fig. 7.7. When the applied field is perpendicular to the magnetisation direction, a net magnetisation can be achieved by effectively tilting the magnetic moments slightly along the field direction. Tilting each spin through a small angle, θ, gives a net magnetisation proportional to $\sin\theta$, that is, the magnetic energy gained through interaction with the field $\sim\theta$. By contrast, when the field is applied parallel to the magnetisation direction, tilting each spin through an angle θ would cost the same amount of energy in terms of disrupting the two sublattices, but now only gains an energy proportional to $(1 - \cos\theta)$ through interaction with the applied field, that is, the energy gained $\sim\theta^2$. This becomes progressively less advantageous as the temperature decreases, so that $\chi_{\parallel} \to 0$ as $T \to 0$, while it can be shown that $\chi_{\perp}$ is approximately constant below T_N. In practice, the anisotropy in susceptibility is only observed in single domain samples, where all spins on a given sublattice point along the same direction. Most macroscopic samples consist of many domains, each with its own magnetisation direction. The low temperature suscepibility is then isotropic, and equal to

$\frac{1}{3}(\chi_\parallel + 2\chi_\perp)$. The susceptibility, therefore, displays a cusp at the Néel temperature, providing a straightforward experimental method to determine the onset of spontaneous antiferromagnetic ordering.

7.6 Ferrimagnetism

In ferrimagnetic materials, we again have two sublattices, but the two are no longer equivalent. The exchange interaction between the sublattices is negative, so that anti-parallel spins are favoured, as in the antiferromagnetic case. We now find $M_A \neq -M_B$, and so there is a net magnetisation below the spontaneous ordering temperature, with $M = M_A + M_B \neq 0$. Whereas most ferromagnets are metals, many ferrimagnets are semiconductors or even insulators. The combination of a very high resisitivity and a net permanent magnetisation means that ferrimagnets such as Fe_3O_4 (referred to as 'ferrite') are particularly useful for a range of devices, including application as high frequency transformers and aerials.

7.7 Spin waves – elementary magnetic excitations

The mean field theory fails at low temperatures for a ferromagnet because it does not correctly predict the energy of low-lying excited states. Consider for instance a closed loop of $2N$ spins, in which each spin is rotated through an angle $\theta = \pi/N$ with respect to the direction of their immediate neighbours, as illustrated in fig. 7.8. This state has a very high energy in mean field theory. Because the average magnetisation M is zero, mean field theory gives $E = 2NE_{ex,i} = 2N(-2Js_i \cdot \langle s_j \rangle) = 0$, as $\langle s_j \rangle = 0$. This is a considerably higher energy than the ground state value, $E_{gs} = -4JNs^2$, in which all the spins are parallel and we assume the magnitude of each spin equals s. We can use the Heisenberg model to evaluate directly the energy of the state shown in fig. 7.8. The interaction energy, W, between

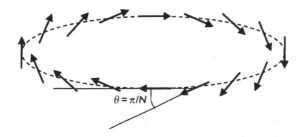

$\theta = \pi/N$

Figure 7.8 A closed loop of 2N spins, for each of which the spin directions is rotated by π/N with respect to its immediate neighbours.

any spin and its two neighbours is given by

$$W = -2Js^2 \cos\theta \tag{7.27}$$

For a sufficiently long loop, θ is small, so we can approximate $\cos\theta$ as $1 - \theta^2/2 = 1 - (\pi/N)^2/2$. The total energy of the chain, $2NW$, is then given by

$$2NW = -4JNs^2 \left(1 - \frac{(\pi/N)^2}{2}\right) = -4JNs^2 + \frac{2Js^2\pi^2}{N} \tag{7.28}$$

and the increase in energy over the ground state, where all spins are parallel, is only $2Js^2\pi^2/N$, which goes to zero as $N \to \infty$. Clearly, mean field theory fails badly in its estimate of the energy of this simple excitation.

The classical unit of spin excitation involves flipping just one spin. (In quantum mechanics, we reduce the total angular momentum by one unit.) We see that we have effectively flipped N spins in order to produce the state illustrated in fig. 7.8. The net energy associated with each individual spin flip, E_{flip} is therefore

$$E_{flip} = \left(\frac{2Js^2\pi^2}{N}\right) \bigg/ N = \frac{2Js^2\pi^2}{N^2} \tag{7.29}$$

We can describe the state in fig. 7.8 as being formed by combining N 'spin waves', each of wavelength $\lambda = 2Na$, where a is the nearest neighbour separation. Each spin wave then has wavevector k, with $k = 2\pi/(2Na) = \pi/Na$, so that $ka = \pi/N$. Substituting for π/N in eq. (7.29), we, therefore, estimate that the energy E_k of the basic quantum unit of spin excitation, referred to as a *magnon*, depends quadratically on the wavevector k, as

$$E_k = \hbar\omega_k \approx 2Js^2(ka)^2 = (2Js^2a^2)k^2 \tag{7.30}$$

We have made a couple of gross assumptions in the above analysis. First, we presumed that the excitation energies add linearly, so that the total energy of N spin waves (eq. (7.28)) is just N times the energy of a single spin wave (hence eq. (7.29)). We then further presumed that the result for a classical spin excitation in eq. (7.29) remains true when we quantise the spins, and so derived an expression as to how the energy of a quantum of spin excitation varies with wavevector k (eq. (7.30)).

It is perhaps surprising to find that the conclusion of eq. (7.30) is in fact true; that is, that the energy E_k of a magnon depends on wavevector k as $E_k \sim k^2$. As a consequence, even at the lowest temperatures, as $T \to 0$, there will always be some long wavelength excitations within an energy kT of the ground state. By contrast, mean field theory overestimates the

elementary excitation energies and, therefore, predicts that the spin alignment approaches perfect ordering more rapidly as $T \to 0$ than is observed
experimentally. We leave it as an exercise to the end of this chapter to show
that in mean field theory the magnetisation M approaches the saturation
value, M_s as

$$M(T) = M_s(1 - e^{-\beta/T}) \qquad (7.31)$$

where β is a material constant. Experimentally, it is found that M
approaches M_s much more slowly, as $M(T) = M_s - \alpha T^{3/2}$. The experimental data can be understood by considering the energetics of magnon
excitation.

Using eqs (7.30) and (5.10) it can be shown that the number of allowed
magnon modes per unit volume with frequency between ω and $\omega + d\omega$ is
given by

$$g(\omega)\, d\omega = \frac{1}{4\pi^2} \left(\frac{\hbar}{2Js^2a^2} \right)^{3/2} \omega^{1/2}\, d\omega \qquad (7.32)$$

where $g(\omega)$ is the magnon density of states. The average number of
magnons, n at temperature T for a mode of frequency ω is given by the
Bose–Einstein distribution function, as

$$n(\omega) = \frac{1}{e^{\hbar\omega/kT} - 1} \qquad (7.33)$$

The number of magnons excited per unit volume, N, is then given by

$$N = \int_0^\infty n(\omega)g(\omega)d\omega$$

$$= C \int_0^\infty \frac{\omega^{1/2}}{e^{\hbar\omega/kT} - 1} d\omega \qquad (7.34)$$

If we make the change of variable $x = \hbar\omega/kT$, the integral in eq. (7.34) can
be rewritten as

$$N = C(kT)^{3/2} \int_0^\infty \frac{x^{1/2}}{e^x - 1} dx = 2.32C(kT)^{3/2} \qquad (7.35)$$

so that the number of magnons excited per unit volume increases as $T^{3/2}$.
As each magnon reduces the overall magnetic moment by $g\mu_B$, the net
magnetisation is indeed found to vary as $M(T) = M_s - \alpha T^{3/2}$, as observed
experimentally.

Mean field theory also breaks down near the Curie temperature, T_c.
We have seen that the susceptibility above T_c is predicted to vary as

$\chi = C/(T - T_c)$. Below T_c the magnetisation M varies in mean field theory as $M(T) \sim (T_c - T)^{1/2}$ (see problem 7.4). Experimentally it is found that χ varies more typically near T_c as $(T - T_c)^{-\gamma}$ and $M(T)$ as $(T_c - T)^{\beta}$, where $\gamma \sim 1.33$ and $\beta \sim 0.33$. The discrepancy again arises because mean field theory fails to take sufficient account of the short range order in a ferromagnet, assuming that because the mean magnetic moment $\langle m \rangle$ is zero, there are effectively no correlations between neighbouring moments. In practice, moments tend to be aligned locally near T_c, but the direction of the local moment fluctuates strongly through the sample.

7.8 Ferromagnetic domains

Despite the existence of spontaneous magnetisation below the Curie temperature, it is well known that ferromagnets (such as steel needles) can apparently lose their magnetisation. This is due to the formation of domains, whereby the magnetisation points in different directions in different parts of the sample. Application of an external magnetic field can cause the domain walls to move, as illustrated in fig. 7.9, leading to a net magnetisation of the sample. This is how a steel needle becomes magnetised when a bar magnet is passed along the needle.

Why are domains formed in a magnetic material? Because they reduce the overall energy of the system. If a sample has only one domain, there will be a large external magnetic field B associated with its macroscopic magnetisation, with the energy stored per unit volume in the external field given by $\frac{1}{2}HB = B^2/2\mu_0$ in free space. With many domains, the external field B is significantly reduced, thereby reducing the overall energy.

Although the boundaries between domains are shown as sharp lines in fig. 7.9, there is in fact a narrow transition region, known as a Bloch wall, between neighbouring domains, across which the magnetisation direction changes smoothly, as illustrated earlier in fig. 7.8, when discussing spin waves. Two competing effects determine the width, w, of the domain

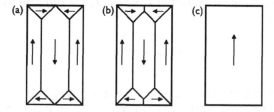

Figure 7.9 (a) Schematic domain structure for a ferromagnet in zero applied field, where the domain pattern is tending to minimise the magnetostatic energy. Application of an external field H can lead first to reversible domain wall motion (b) and then (c) to irreversible elimination of domains.

wall, namely spin wave energy, which we have already discussed, and anisotropy energy, which we now consider.

In general, magnetic moments tend to line up along high symmetry directions in a crystal, giving what are referred to as 'easy' and 'hard' directions of magnetisation. This means that within a crystal there are preferred directions of magnetisation, when the atomic moments are pointing in particular crystallographic directions. It requires work, associated with changes in the electron charge distribution, to rotate the moments to other directions. It is usual to describe this magnetocrystalline anisotropy energy, U_c, per unit volume in terms of empirical anisotropy coefficients, K. For cubic crystals such as Fe and Ni, we can write

$$U_c = K_1(\alpha_1^2\alpha_2^2 + \alpha_2^2\alpha_3^2 + \alpha_3^2\alpha_1^2) + K_2\alpha_1^2\alpha_2^2\alpha_3^2 \qquad (7.36a)$$

where α_1, α_2 and α_3 are the direction cosines of the spin direction relative to the three cube axes. For a hexagonal close-packed crystal such as Co, where the easy axis is along the z-direction, we can write

$$U_c = K_1 \sin^2\theta + K_2 \sin^4\theta \qquad (7.36b)$$

where θ is the angle between the spin-direction and the z-axis.

The exchange interaction favours neighbouring spins s_1 and s_2 to be parallel. The spin wave contribution to the domain wall energy, therefore, decreases as the spin wavelength increases and the wall gets wider. By contrast, the anisotropy energy favours narrow walls, so that as few spins as possible are directed away from the easy direction(s).

Consider for example a simple cubic crystal with lattice constant a, where a domain wall is N layers thick, with the angle changing by $\theta = \pi/N$ between successive layers: the spin direction is therefore reversing, changing by π across the domain wall. The increase in exchange energy for a single line of spins across the domain wall is then equal, from eq. (7.28), to $Js^2\pi^2/N$, where J is the nearest neighbour exchange interaction. The increase in exchange energy, ΔE_{ex}, per unit area of domain wall is then given by

$$\Delta E_{ex} = \frac{Js^2\pi^2}{Na^2} = \frac{E_{ex}\pi^2 a}{6N} \qquad (7.37)$$

where we have used the fact that the exchange energy per unit volume, E_{ex}, equals $6Js^2/a^3$ in a perfectly ordered simple cubic ferromagnet.

The total anisotropy energy E_{an} per unit area for a wall of width $w = Na$ is given approximately as

$$E_{an} = K(Na) \qquad (7.38)$$

where K is a mean or average anisotropy constant. The Bloch wall energy per unit area, $E_{\text{Bloch}} = \Delta E_{\text{ex}} + E_{\text{an}}$, is then minimised when

$$\frac{\partial E_{\text{Bloch}}}{\partial N} = \frac{\partial}{\partial N}\left(\frac{E_{\text{ex}}\pi^2 a}{6N} + KNa\right) = 0 \tag{7.39}$$

that is, when

$$w = Na = \pi a\sqrt{\frac{E_{\text{ex}}}{6K}} \tag{7.40}$$

As the exchange energy, E_{ex}, is generally several orders of magnitude larger than the anisotropy energy, K, we conclude that each Bloch wall will be many layers thick: substituting appropriate values for Fe, one finds that $N \approx 350$, with the wall thickness then ~ 1000 Å.

The total Bloch wall energy per unit area can be estimated by substituting the value of $w = Na$ from eq. (7.40) into eq. (7.39), giving

$$E_{\text{Bloch}} = \pi a\sqrt{\frac{2E_{\text{ex}}K}{3}} \tag{7.41}$$

For a spherical particle of radius R, the energy to create a single Bloch wall then scales as R^2, whereas the energy stored in the external field scales as R^3. Below a critical radius R_c, the energy to create a Bloch wall therefore exceeds that stored in the external magnetic field, so that single domain particles become preferred.

7.9   High-performance permanent magnets

The unique attraction of permanent magnets is that (barring the eventual development of a room temperature superconductor) they provide magnetic flux with no continuing expenditure of energy. Indeed their ferromagnetism has its origin in resistanceless electric currents circulating on the atomic scale. Furthermore, their ability to generate complex field patterns with intense spatial variations is unrivalled by any electromagnetic device. The surface current needed to generate a flux pattern similar to that of a long cylindrical magnet with $\mu_0 M = 1\,\text{T}$ is 796 kA/m. Solenoids, whether resisitive or superconducting, able to produce such fields would have to be several centimetres in diameter to accommodate the necessary ampere-turns.

Intrinsic magnetic properties set the limits on the potential development of any particular material. These properties include the magnitudes of the saturation magnetisation, M_s, and of the magnetocrystalline anisotropy energy, U_c, and the existence or otherwise of a suitable easy direction. These

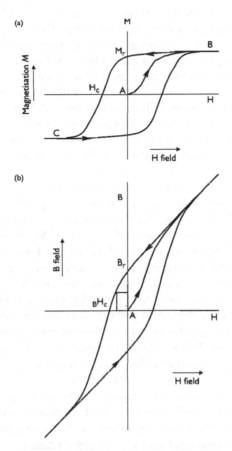

Figure 7.10 Hysteresis loops for a permanent magnet (a) $M(H)$ loop and (b) $B(H)$ loop. Virgin curves start at the origin (A). Key characteristics of the magnet include: (i) its remanence magnetisation (M_r at $H = 0$), (ii) coercivity fields, H_c and $_BH_c$, the reverse fields where M and B, respectively, return to zero, and (iii) figure of merit, $(-BH)_{max}$, the shaded area in (b).

aspects can be predicted from suitable theoretical analysis. In addition, details of the metallurgical microstructure determine the ease or otherwise with which domain walls will move in an applied magnetic field, and hence ultimately determine how effective a particular sample can be in maintaining a strong permanent magnetisation. With nanometre-sized grains, for instance, each grain could act as a single domain, eliminating the possibility of reversible domain wall motion (Davies 1994).

An external magnetic field usually changes the state of magnetisation of a ferromagnet in a way that is nonlinear and irreversible (see fig. 7.9). A typical magnetisation curve $M(H)$ is illustrated in fig. 7.10(a). Starting at the zero field (A), an initially demagnetised sample is subject to an increasing magnetic field. The magnetisation increases with field until the sample becomes fully magnetised, with magnetisation M_s (B). When the field is removed, the magnetisation decreases to a value M_r, referred to as the *remanence* magnetisation. When a field H is now applied in the opposite direction the magnetisation will return to zero at a finite field value, H_c, referred to as the *coercivity* field. With further increase in the magnitude of the field, the sample again approaches the saturation magnetisation value, M_s (C). M_s is an intrinsic property of the ferromagnetic phase but the remanence M_r and coercivity H_c are not.

The $B(H)$ curve of fig. 7.10(b) is related to the $M(H)$ curve of fig. 7.10(a) by eq. (6.5), $B(H) = \mu_0(H + M(H))$. The remanence B_r in this case is just $\mu_0 M_r$, but the magnitude of the coercivity on the B–H plot of fig. 7.10(b), $_B H_c$, is smaller than in fig. 7.10(a), because $B = \mu_0(H + M)$ goes to zero when M is still >0. Coey (1996) remarks that $M(H)$ curves tend to be measured by physicists interested in magnetisation, whereas $B(H)$ curves are of more interest to engineers. This is because we can use the $B(H)$ curve of fig. 7.10(b) to determine the maximum potential energy which can be stored in a magnet and hence design its performance for specific applications.

It can be shown that the maximum potential energy density is obtained (at least for an ellipsoidal sample) at the point where the product $-\frac{1}{2}B \cdot H$ is maximised. The figure of merit $(BH)_{max}$, corresponding to the shaded area in fig. 7.10(b), is equal to twice this maximum potential energy (Coey 1996; Myers 1997). We can estimate the upper limits for the performance of any magnet by noting that the figure of merit is maximised when the remanence has its largest possible value, $M_r = M_s$, and the magnetisation remains independent of reverse field up to H_c, in which case $(BH)_{max} = (\frac{1}{2}\mu_0 M_s) \cdot (\frac{1}{2}M_s)$, giving the following intrinsic limitations on the performance of permanent magnets:

$$B_r \leq \mu_0 M_s \tag{7.41a}$$

$$_B H_c \leq M_s \tag{7.41b}$$

and

$$(BH)_{max} \leq \frac{\mu_0 M_s^2}{4} \tag{7.41c}$$

The major considerations in the design of high-performance magnets are, therefore, to choose a ferromagnet with a large value of M_s and a high coercivity. As stated earlier, this depends both on the alloys used and also on

fine details of the microstructure. The largest magnetic moments are generally associated with the so-called rare-earth series of elements, which have
a partially filled 4f or 5f shell of electrons (see the Periodic Table in fig. B.1).
Because this f-shell has a small radius compared to the element's valence
electron radius it can retain its total spin and orbital angular momenta in
a solid environment, giving a large magnetic moment. The small radius,
however, also means that the moment is less sensitive to its local environment, implying smaller values of the ferromagnetic coupling coefficients
and magnetocrystalline anisotropy energy, U_c. This then limits the coercivity values, $_BH_c$. By contrast, the ferromagnetic transition metal elements
(Fe, Co, and Ni) have a partially filled 3d shell, whose radius is comparable
to that of the 4s valence electrons. This explains the observation in Chapter
6 that the orbital angular momentum is quenched ($L = 0$) when these elements are incorporated in a solid. But the larger 3d shell radius also brings
the benefit of larger values of the ferromagnetic coupling coefficients and
anisotropy energy.

Figure 7.11 shows how the maximum energy product $(BH)_{max}$ has
improved over the past century, as the values of M_r and $_BH_c$ have
increased in successive generations of ferromagnets. The energy product
has increased exponentially, doubling every twelve years. Alloys mainly
composed of iron have $\mu_0 M_s$ less than about 2 T, which from eq. (7.41c)
sets an upper limit on $(BH)_{max}$ of order 800 kJ/m^3, placing a ceiling on

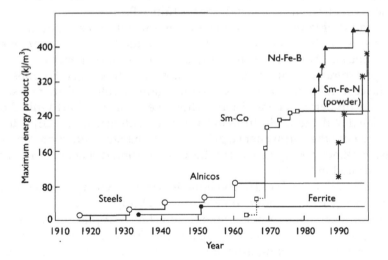

Figure 7.11 Progress in improving the figure of merit $(BH)_{max}$ during the twentieth
century. The energy product increased approximately exponentially, doubling every twelve years (from Skomski and Coey, 1999, by permission of
IoP Publishing).

the progress shown in fig. 7.10. The heavy rare-earths Dy and Ho have the largest atomic moment $(10\,\mu_B)$ of any element in the periodic table and so could in theory show energy products as high as $3000\,kJ/m^3$, enabling a flux density approaching $4\,T$. However, prospects are poor for realising room temperature ferromagnetic alloys based on these elements. They are likely to be restricted to cryogenic applications, if they are ever used at all. Nevertheless, the progress shown in fig. 7.10 has enabled the development of a multi-billion dollar industry and the widespread use of permanent magnets in applications such as motors and generators, actuators and printers, loudspeakers and magnetic resonance imaging devices, as well as a diverse range of scientific instruments from large-scale colliders to tiny ammeters. A considerably more detailed discussion of the science and application of high-performance magnets is given in Coey (1996).

7.10 Itinerant ferromagnetism

We have assumed throughout this chapter that the magnetisation in a ferromagnet is due to interactions between localised magnetic moments on neighbouring ions. There are, in addition, several metals, in particular iron and nickel, whose ferromagnetism is due to interactions between delocalised, so-called itinerant conduction electrons. This is entirely analogous to the situation considered in the previous chapter, where we saw that paramagnetism can be associated not just with isolated spins but also with delocalised electrons in a metal (see Section 6.8).

We can divide the conduction band in a metal into spin-up and spin-down sub-bands. Potential energy is gained due to the exchange interaction if one sub-band is preferentially occupied at the expense of the other. This preferential occupation, however, costs kinetic energy. For most metals the cost in kinetic energy exceeds the gain in potential energy, so that they remain paramagnetic. There is, however, a narrow partly-filled band associated with the 3d electrons in Ni and Fe, leading to a very large density of states near the Fermi energy, $g(E_F)$. The exchange interaction can dominate over kinetic energy effects in this case, leading to a net magnetisation, as illustrated in fig. 7.12.

The dependence of band ferromagnetism on the density of states near the Fermi energy can be understood as follows. Suppose a d band possesses equal numbers of spin-up and spin-down states. Application of an external magnetic field H can lead to a relative shift, ΔE, in the two sub-bands, with electrons transfering from spin-up to spin-down states, as in fig. 6.7. If we now allow an exchange interaction between electrons of the same spin, then the shift in the two sub-bands, $\Delta E = \mu_0\mu_B H$ is increased by an amount $J\Delta N$ relative to the purely paramagnetic case. $\Delta N = (N_+ - N_-)$ is the difference in occupation of the two sub-bands, and J is an effective exchange energy between parallel spins. We can, therefore, write the

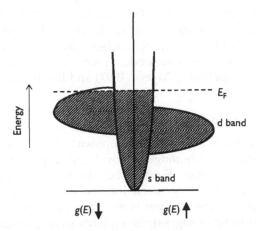

Figure 7.12 Schematic density of states of ferromagnetic Ni, showing the density of spin-up states as a function of energy on the left-hand side, and spin-down states on the right-hand side. There is a large density of states near the Fermi energy, E_F, associated with the Ni 3d electrons. The exchange interaction shifts the spin-up and spin-down states with respect to each other, leading here to a preferential occupation of the spin-up band and an overall net gain in energy. The wider band in the figure, with a much lower density of states is due predominantly to interactions between 4s states on the Ni atoms.

shift ΔE as

$$\Delta E = 2\mu_0\mu_B H + J\Delta N \tag{7.42}$$

Since all the changes occur near the Fermi energy, we can relate ΔN to ΔE as

$$\Delta N = g(E_F)\Delta E/2 \tag{7.43}$$

where $g(E_F)$ is the total density of states at the Fermi energy. We now substitute eq. (7.42) in (7.43) to determine the induced magnetisation $M = \mu_B\Delta N$ as

$$M = \frac{1}{2}\mu_B g(E_F)\left(2\mu_0\mu_B H + \frac{JM}{\mu_B}\right) \tag{7.44}$$

We can re-arrange eq. (7.44) as we did for eqs (7.11) and (7.21) to find that $M = \chi H$, with the susceptibility χ given by

$$\chi = \frac{\mu_0\mu_B^2 g(E_F)}{1 - \frac{1}{2}Jg(E_F)} \tag{7.45}$$

We have already seen that spontaneous magnetisation arises when the susceptibility diverges ($\chi \to \infty$). This occurs here when $\frac{1}{2}Jg(E_F) > 1$. This relation, known as the Stoner criterion, describes the key requirement to achieve ferromagnetism due to delocalised electrons in a metal. Further details and analysis of the band model of ferromagnetism can be found in several other texts, including Myers (1997) and Ibach and Lüth (1995).

7.11 Giant magnetoresistance

We saw in Chapter 5 how the development of epitaxial growth techniques has enabled a wide range of novel and improved semiconductor structures and applications. Similar techniques can be applied to grow magnetic multilayers. The greatest interest has focussed on structures where a thin ferromagnetic layer is separated by a non-magnetic layer from a second thin ferromagnetic layer. Such structures can demonstrate giant magnetoresistance, an effect with potentially significant applications.

It can be shown that when the indirect exchange interaction between two magnetic ions is mediated by free conduction electrons then the value of the exchange interaction will decay with separation distance, r, as $(1/r^3)\cos(2k_Fr)$, where k_F is the wavevector at the Fermi energy. Details of this interaction, which is associated with the names of Ruderman, Kittel, Kasuya, and Yosida (RKKY), can be found elsewhere (Kittel 1968). As a consequence of the RKKY interaction, the sign of the exchange interaction between two thin ferromagnetic layers oscillates as the thickness of the intermediate non-magnetic layer is increased.

Consider a thin ferromagnetic layer where the magnetisation vector M points in the plane of the layer. It is possible by a suitable choice of

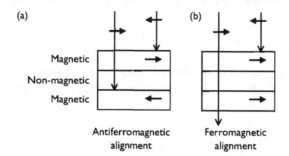

Figure 7.13 (a) The ground-state magnetisation in a magnetic multilayer where anti-ferromagnetic coupling is favoured between neighbouring magnetic layers. (b) Application of a sufficiently strong external magnetic field aligns the magnetisation in neighbouring layers. GMR is explained by spin-dependent scattering of conduction electrons passing through the structure: both spin directions are scattered in case (a), whereas only one is scattered in case (b), resulting in a lower overall resistance in (b).

intermediate layer thickness to grow a multilayer structure where the inter-action between neighbouring magnetic layers is antiferromagnetic, with the magnetisation then given by $-M$ in the second layer, as illustrated in fig. 7.13(a). If a sufficiently large external magnetic field is applied within the plane, say of order 0.5 T, then the magnetisation vectors in the two layers will align with each other, as illustrated in fig. 7.13(b). Remarkably, it is observed that when current is then driven along the growth direction, the resistance of the multilayer structure can decrease by a factor of order two between fig. 7.13(a) and (b). This exceptionally large variation of resistance with applied magnetic field is referred to as giant magnetoresistance (GMR).

A qualitative explanation of GMR can be obtained by considering the schematic density of states in fig. 7.12, where the spin-up d band lies below the Fermi energy and is completely filled, while the spin-down d band cuts the Fermi energy, and is therefore only partly filled. The current in such a ferromagnet is carried predominantly by the s electrons near the Fermi energy. Because of the large density of unfilled spin-down d states, a spin-down s electron can experience strong s–d scattering, limiting its conductance. By contrast with all spin-up d states filled, a spin-up s electron will experience no s–d scattering, and will therefore have a higher conduc-tivity. In an antiferromagnetically aligned multilayer structure (fig. 7.13(a)) spin-up electrons will then be scattered by one layer and spin-down elec-trons by the second layer, so that both spin-up and spin-down electrons have a high resistance. By contrast, when the interaction is ferromag-netic, only one type of spin, say spin-down, is scattered by the multilayer (fig. 7.13(b)); and the low resistance spin-up channel then effectively shorts out the high resistance spin-down channel, leading to the large observed decrease in resistivity.

The occurence of GMR has excited considerable interest, because it offers the means to develop very responsive magnetoresistive sensors and pos-sibly even a magnetic transistor based on magneto-resistive effects (see Barthélémy et al. 1994). The main requirement to achieve a practical GMR device is to reduce the magnetic field at which complete switching to ferro-magnetic inter-layer alignment occurs, in order to detect the weak magnetic fields associated with very small data storage elements.

GMR has been displayed in a wide range of magnetic multi-layer struc-tures and considerable progress has been made both in understanding the underlying mechanisms, and in developing materials which show sub-stantial MR changes in modest fields. Interesting effects have also been observed by combining ferromagnetic metals and semiconductors in a sin-gle nanostructure. For example, when spin polarised electrons are excited in a semiconductor by circularly polarised light, the tunnelling current into a neighbouring ferromagnetic layer can depend on the orientation of the magnetisation and consequent density of empty states at the ferromag-net Fermi energy. This effect can be applied to the imaging of magnetic domains. Other interesting effects are expected if spin-polarised electrons can be injected from a ferromagnetic metal into a superconductor. There is now considerable interest in this emerging field, referred to as 'spintronics'. Considerable progress has been made through the application of quantum theory to understand magnetic materials, which has led to the development of materials with a wide diversity of properties, some of which we considered in this and the previous chapter. Nevertheless, there still remain many fundamental problems to explore. With so many promising applications waiting for the improved integration of information storage and logic operations in microelectronics, it can be expected that research in magnetic nanostructures will continue to be of considerable importance and interest.

References

Barthélémy, A. *et al.* (1994) *Physics World* **7(11)** 34.

Coey, J. M. D. (1996) (ed.) *Rare Earth Iron Permanent Magnets*, Clarendon Press, Oxford.

Davies, H. (1994) *Physics World* **7(11)** 40.

Ibach, H. and H. Lüth (1995) *Solid State Physics, An Introduction to Principles of Materials Science*, Springer-Verlag, Berlin.

Kittel, C. (1968) *Solid State Physics* **23** 1.

Myers, H. P. (1997) *Introductory Solid State Physics*, 2nd edn, Taylor and Francis, London.

Skomski R. and J. M. D. Coey (1999) *Permanent Magnetism*, IoP Publishing, Bristol.

Trapp C. and J. W. Stout (1963) *Phys. Rev. Lett.* **10** 157.

American Institute of Physics Handbook, 3rd edn, McGraw-Hill, New York (1972).

Problems

7.1   Consider a spin-$\frac{1}{2}$ ferromagnet for which we can write the spontaneous magnetisation $M = N\mu_B \tanh(x)$ with $x = \mu_0 \mu_B \lambda M / kT$. Show that the Curie temperature T_c is given by $kT_c = N\mu_0\mu_B^2\lambda$. Determine at what fraction of the Curie temperature the spontaneous magnetisation drops to 80 per cent of its value at $T = 0$, and then to 50 per cent of its value at $T = 0$.

7.2   A ferromagnetic oxide has a transition temperature of 150 K. If the magnetic carriers have a magnetic moment of μ_B and spin $S = \frac{1}{2}$, determine the ratio of the magnetisation at 300 K to that at 0 K in a field of 0.1 T. You may assume the effective field model outlined in Sections 7.4 and 7.5 to be appropriate for this case.

7.3   The anisotropy energy U_c in a ferro- (or ferri-) magnet of cubic symmetry is given by eq. (7.36a) as

$$U_c = K_1(\alpha_1^2\alpha_2^2 + \alpha_2^2\alpha_3^2 + \alpha_3^2\alpha_1^2) + K_2\alpha_1^2\alpha_2^2\alpha_3^2 \qquad (7.36a)$$

Derive expressions for $(W_{(110)} - W_{(100)})$ and $(W_{(111)} - W_{(100)})$, where

$$W = \int_0^{M_s} H \cdot dM$$

is the work done in magnetising a single crystal to saturation in the directions indicated.

7.4 Consider a ferromagnet in which there are N magnetic ions per unit volume, each with a magnetic moment of μ_B and spin $S = \frac{1}{2}$, with $g = 2$. Use mean field theory and estimate the value of the Brillouin function $B_{1/2}(x)$ at large x to show that the magetisation M varies with temperature T at low temperatures as

$$M(T) = M_s(1 - e^{-\beta/T})$$

relating the value of β to the magnetic moment and Curie temperature. Show also by expanding $B_{1/2}(x)$ for small x that the magetisation using mean field theory varies near the Curie temperature as $(T_c - T)^{1/2}$.

7.5 Show that if we use the classical (Langevin) theory of paramagnetism, then the magnetisation $M(T)$ in a ferromagnet will vary near $T = 0$ as $M(T) = M_s - \alpha T$. Explain why this classical mean field result is closer to the experimentally observed behaviour than the quantised mean field result of problem 7.4.

7.6 The unit cell of $NiFe_2O_4$ has eight Ni^{2+} ions on B sites and sixteen Fe^{3+} ions equally distributed between A and B sites. The dominant exchange interaction is a negative (antiferromagnetic) AB interaction. Orbital angular momentum is assumed to be quenched on the Ni^{2+} ions. Show that in the fully ordered state at low temperature, the magnetic moment per unit cell is $16 \, \mu_B$. What would be the magnetic moment per unit cell in $Zn_{0.25}Ni_{0.75}Fe_2O_4$? The outer electronic configuration of Zn^{2+} is $3d^{10}$, and it is found experimentally that the Zn^{2+} ions go on to A sites, displacing Fe^{3+} ions on to B sites.

7.7 A ferrimagnetic crystal consists of two sublattices, A and B. The effective fields F_A and F_B associated with the two sublattices are given by eq. (7.18), $F_{A(B)} = H_0 - \lambda M_{B(A)}$, while at high temperatures we have

$$M_A = \frac{1}{2} \frac{C_A}{T} F_A \qquad M_B = \frac{1}{2} \frac{C_B}{T} F_B$$

Determine the high temperature susceptibility of the crystal, and show that the critical temperature, T_c, below which spontaneous magnetisation occurs is given by $T_c = \lambda(C_A C_B)^{1/2}/2$. By comparing your result with eq. (7.13) for a ferromagnet deduce how the high temperature susceptibility can be used to distinguish between ferrimagnetic and ferromagnetic crystals.

Chapter 8

Superconductivity

8.1 Introduction

Ever since its discovery in 1911, superconductivity has held out tantalising possibilities for widespread application, including in particular highly efficient electricity distribution, and also high field magnets. The field of superconductivity poses several significant challenges. The first of these is to understand the mechanisms which give rise to the phenomenon. This cannot be done in the one-electron approximation used so far in this book, which assumes each electron effectively acts independently of the others. The difficulty in deducing the mechanism of superconductivity is emphasised by the fact that it took nearly fifty years from its first observation to develop a satisfactory theory of the underlying processes. Guided by this theory, a series of improved superconductors were developed through the 1960s and 1970s, but the highest critical temperature, T_c, at which superconductivity could be observed remained stubbornly low – about 20 K – severely limiting the potential applications. Many believed, with theoretical justification, that this was close to the maximum achievable T_c. The discovery by Bednorz and Müller (1986) of a superconducting compound, $La_{2-x}Ba_xCuO_4$ with $T_c = 35$ K has opened up a new and exciting phase in superconductivity research, leading to the development of new materials with critical temperatures above liquid nitrogen temperature (77 K), but also posing the question as to the mechanism underpinning this high-temperature (high-T_c) superconductivity. At the time of writing (2001), this question remains unresolved: the key differences between high-T_c and conventional superconductors have been highlighted, but the challenge remains to establish a single, widely-accepted model of the high-T_c superconducting mechanism.

We begin this chapter with a review of some of the key experimental features of superconductivity, and then outline the BCS theory of conventional superconductivity (Bardeen, Cooper and Schrieffer 1957), and its consequences. The chapter and book then conclude with an overview of current progress in high-T_c superconductivity.

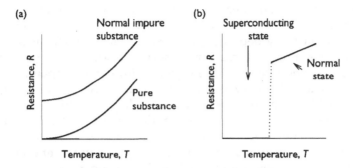

Figure 8.1 Schematic illustration of (a) the variation of resistivity at low temperature in a normal impure metal, and in a highly purified crystal of the same metal; compared with (b) the superconducting transition at low temperature in mercury.

8.2 Occurence of superconductivity

We consider first a normal metal, which has a DC electrical resistivity, ρ, so that the current density J in an applied electric field E is given by

$$E = \rho J \qquad (8.1)$$

The resistance in such a metal has two main components, due to (i) thermal vibrations of the atoms, and (ii) impurities and defects in the crystal. Both these processes scatter electrons. (Electrons are *not* scattered by a perfect lattice.) As the temperature is decreased, the lattice vibrations decrease, and it is consequently observed that the resisitivity also decreases. Figure 8.1(a) shows schematically the resisitivity of a normal, impure metal, and of a highly purified single crystal of the same metal, near $T = 0\,\mathrm{K}$: the resistivity remains finite in the former case, but $\rho \to 0$ as $T \to 0$ in the pure sample.

In 1911, the Dutch scientist, Heike Kamerlingh Onnes, measured the resistance of mercury at low temperature, three years after he had first succeeded in the liquefaction of helium. He found that mercury underwent a dramatic transition at a finite critical temperature ($T_c \sim 4.2\,\mathrm{K}$) from a normal to a superconducting state, with $\rho \sim 0$ below T_c (fig. 8.1(b)). He was unclear because of experimental uncertainties whether $\rho \sim 0$ or $\rho \equiv 0$ exactly. The best means to determine a low resistance R accurately is to observe current flow in a closed loop of self-inductance, L, where the current should decay with a time constant, $\tau = L/R$. By observing the persistent current for a year, File and Mills (1963) placed a lower limit on τ in a superconducting closed loop of 100 000 years, requiring a superconducting resistivity $\rho < 10^{-26}\,\Omega\,\mathrm{m}$, over 10^{15} times smaller than in the normal state, thereby justifying the assumption that $\rho \equiv 0$ in the superconducting state.

Many (but not all) metals are superconductors, with the observed transition temperature ranging from $T_c = 0.01\,K$ in tungsten to $T_c = 9.2\,K$ for niobium (Nb). It is perhaps surprising that the establishment of superconductivity depends only weakly on the material purity, nor is any change in crystal structure observed below the superconducting transition temperature. A wide range of alloys and compounds are also superconductors. Nb_3Ge was found in 1972 to have $T_c = 23\,K$, and this was justifiably believed to be close to the highest achievable transition temperature in a conventional superconductor. Figure 8.2 shows that following the discovery in 1986 of $La_{2-x}Ba_xCuO_4$ with $T_c = 35\,K$, there was a rapid demonstration, in the following two years, of other, higher temperature superconductors, all based on copper oxide compounds, including $Tl_2Ba_2Ca_2Cu_3O_{10}$, with $T_c = 125\,K$. This is still close to the maximum T_c observed in any superconducting material. More recently, the field of conventional superconductivity has been re-invigorated with

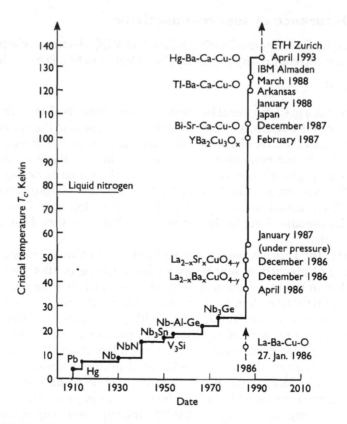

Figure 8.2 Transition temperature of a selection of superconductors plotted against their year of discovery (after Bednorz and Müller 1988).

the serendipitous discovery that MgB_2 has a T_c value approaching 40 K (Nagamatsu *et al.* 2001).

8.3 Magnetic behaviour and Meissner effect

Superconductors are not just 'perfect conductors'. This is shown by their behaviour in a magnetic field, which differs from that which would be predicted for a perfect conductor. We require that the electric field, $E = 0$ inside a perfect conductor, as charge can re-distribute instantaneously to cancel any non-zero field in the conductor. From Maxwell's equations, we therefore have

$$-\frac{\partial B}{\partial t} = \nabla \times E = 0 \tag{8.2}$$

in a perfect conductor, so that the magnetic field B must be time-independent inside such a perfect conductor.

Consider a conductor with finite resistance ($T > T_c$) placed in a finite applied magnetic field, B_{app}. We saw in Chapter 6 that the magnetic susceptibility of such a conductor is generally very small ($\chi \sim 10^{-5}$) so the magnetic field effectively penetrates the conductor, and we can say $B = B_{app}$ inside the conductor (fig. 8.3(i)). We now reduce the temperature below the superconducting critical temperature ($T < T_c$). If the superconductor were just a perfect conductor, then the magnetic field B inside the material should remain unchanged, with $B = B_{app}$ (fig. 8.3(iia)). This, however, is *not* what happens. Instead, a current is induced on the superconductor surface, whose effect is to expel the magnetic flux from within the superconductor, so that $B = 0$ inside the superconductor, and B consequently shows a measurable increase outside the superconductor (fig. 8.3(iib)). If we now remove the applied field with $T < T_c$, we would expect an induced surface current in the perfect conductor to maintain $B = B_{app}$, leading to a measurable value of B outside the perfect conductor (fig. 8.3(iiia)). Instead, we find $B = 0$ both inside and outside the superconductor (fig. 8.3(iiib)).

The superconductor is therefore more than a perfect conductor. It also behaves like a *perfect diamagnetic* material. This is referred to as the Meissner effect. Inside the superconductor, the magnetic field B is given by

$$B = \mu_0(H_{ext} + M) = 0 \tag{8.3}$$

where M is the magnetisation due to the induced surface currents and H_{ext} the externally applied field, so that $M = -H_{ext}$, and

$$\chi = M/H = -1 \tag{8.4}$$

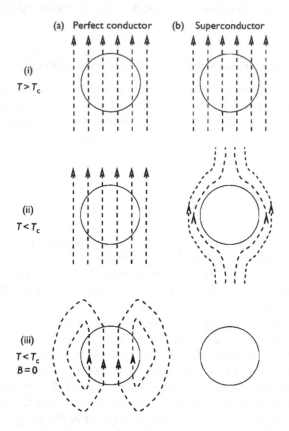

Figure 8.3 Comparison of the magnetic behaviour of (a) a perfect conductor and (b) a superconductor. (i) Above a critical temperature, T_c, both are in the normal state, and an applied static magnetic field penetrates the metal. (ii) On cooling below T_c the magnetic field remains unchanged inside the perfect conductor, but is screened out inside the superconductor, due to surface currents induced below T_c. (iii) When the applied field is turned off, surface currents would be induced to maintain finite B inside the perfect conductor. By contrast, the field B is then zero inside and outside the superconductor.

8.4   Type I and Type II superconductors

The response of a superconductor to an increasing external field, H_{ext}, can be divided into two broad categories, referred to as Type I and Type II superconductors. In a Type I superconductor, the magnetic flux suddenly penetrates into the superconductor at a critical field, $H = H_c$, above which field the material reverts from the superconducting to the normal state. For a pure material this is a reversible transition, with the magnetisation then

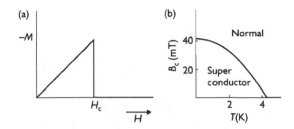

Figure 8.4 (a) Induced magnetisation, M, as a function of applied field, H, in a Type I superconductor. At $H = H_c$, the superconductor reverts to the normal state, so that above H_c, $M \sim 0$. (b) Variation of critical field $B_c = \mu_0 H_c$ with temperature for mercury.

behaving as in fig. 8.4(a). For a type I superconductor, the critical field at 0 K is typically $\sim$ 10–100 mT, and varies with temperature (fig. 8.4(b)) as

$$H_c(T) = H_c(0)\left[1 - \left(\frac{T}{T_c}\right)^2\right] \tag{8.5}$$

These low values of H_c are unfortunate, because they eliminate the possibility of applying Type I superconductors in high-field magnets.

Superconducting magnets are made from Type II superconductors, which are generally made of alloys rather than elements. A Type II superconductor expels all flux, and is perfectly diamagnetic up to a critical field, H_{c1} (fig. 8.5(a)). Above H_{c1}, there is partial penetration of magnetic flux through the metal, and the magnitude of the magnetisation, M, then decreases with increasing field until an upper critical field, H_{c2}, beyond which the material reverts fully to the normal state. The upper critical field, H_{c2}, can be of order 100 T, three orders of magnitude greater than in a Type I superconductor. Between H_{c1} and H_{c2}, the material is in a mixed state: the magnetic field penetrates through thin cylindrical normal regions. These are referred to as flux vortices. Each vortex acts like a solenoid which encloses a single quantum of magnetic flux, $\Phi_0 = h/2e$; the current flowing around the edges of the vortex allows the field to penetrate the vortex region, while leaving B (and ρ)= 0 in the superconducting regions between the vortices (fig. 8.5(b)). The current flow in a pure Type II material is dissipative. When an electric current of current density, j, flows through the material, each vortex (and hence the normal regions) experience a net force, $F = j \times \Phi_0$, leading to normal resistive current flow. Considerable effort has therefore been devoted to developing imperfect materials, where the flux lines are pinned by impurities such as dislocations and grain boundaries, thereby eliminating the resistive current associated with the movement of the normal vortex regions through the material. We shall return later

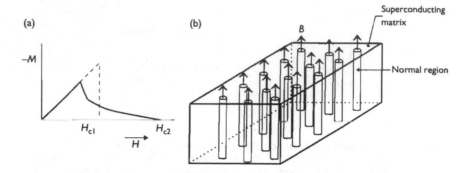

Figure 8.5 (a) Induced magnetisation, *M*, as a function of applied field, *H*, in a Type II superconductor. Above H_{c1}, there is partial penetration of magnetic flux into the superconductor. Above $H = H_{c2}$, the superconductor reverts fully to the normal state. (b) Between H_{c1} and H_{c2}, the magnetic flux penetrates through thin cylindrical normal regions, each of which encloses a single quantum of magnetic flux, and is refered to as a flux vortex. (From P. J. Ford and G. A. Saunders (1997) *Contemporary Physics* **38** 75 © Taylor & Francis.)

to consider what determines whether a superconductor will be Type I or Type II, but turn first to seek an explanation of the Meissner effect.

8.5 Electromagnetic momentum and the London equation

To explain the Meissner effect, we first introduce the concept of *electromagnetic momentum* for a classical charged particle. This allows us to deal more efficiently and elegantly with the motion of particles in an applied magnetic field, *B*. Because the divergence of the magnetic field is zero, $\nabla \cdot B = 0$, we can always define a vector potential, *A*, such that $B = \nabla \times A$. If we consider a particle of mass *M* and charge *q* moving with velocity *v* in the applied field $B = \nabla \times A$, then the total momentum, *p*, of the particle can be defined as

$$p = Mv + qA \tag{8.6}$$

where *Mv* is the kinetic momentum, and *qA* is referred to as the electromagnetic momentum.

It can be shown that the total momentum *p* is conserved in the presence of time-dependent magnetic fields, even when *B* is constant at the position of the charged particle. The general proof is beyond the scope of this text, but the conservation of total momentum can be readily demonstrated by a specific example.

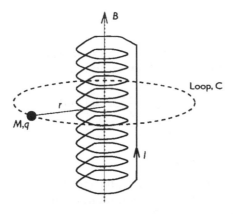

Figure 8.6 A charged particle of mass M and charge q sitting outside a solenoid expe-
        riences a net acceleration when the current through and field inside the
        solenoid decay to zero. This is despite the fact that B = 0 at all times
        outside the solenoid.

Consider a stationary particle of mass M and charge q at a distance r from the centre of a long (effectively infinite) superconducting solenoid, with current I at a temperature $T < T_c$ (see fig. 8.6). We note that although $\mathbf{B}$ is finite inside the solenoid, $\mathbf{B} = 0$ on the loop, C, where the charged particle is positioned outside the solenoid. If we heat the solenoid above T_c then the current I will decay. Because the field $\mathbf{B} \rightarrow 0$ inside the solenoid, the magnetic flux Φ through the loop C will decrease, leading to an induced e.m.f. (electromotive force) around the loop C, given by Faraday's Law as

$$\oint_C \mathbf{E} \cdot d\mathbf{l} = -\frac{d\Phi}{dt} \tag{8.7}$$

where the flux Φ through the loop is given by

$$\Phi = \int_S \mathbf{B} \cdot d\mathbf{S} = \int_S (\nabla \times \mathbf{A}) \cdot d\mathbf{S} \tag{8.8}$$

or, using Stokes' theorem,

$$\Phi = \oint_C \mathbf{A} \cdot d\mathbf{l} \tag{8.9}$$

Substituting (8.9) in (8.7), we then find

$$\oint_C \mathbf{E} \cdot d\mathbf{l} = -\oint_C \frac{\partial \mathbf{A}}{\partial t} \cdot d\mathbf{l} \tag{8.10}$$

and we can choose A so that

$$E = -\frac{\partial A}{\partial t}$$ (8.11)

The change in kinetic momentum of the particle between times t_1 and t_2 is the impulse of the force $F = qE$ acting on it during this interval:

$$[Mv]_{t_1}^{t_2} = \int_{t_1}^{t_2} qE \, dt = -q \int_{t_1}^{t_2} \frac{\partial A}{\partial t} dt = [-qA]_{t_1}^{t_2}$$ (8.12)

Rearranging eq. (8.12) we find

$$Mv_1 + qA_1 = Mv_2 + qA_2$$ (8.13)

so that $p = Mv + qA$ is conserved at all times t, as the current decays. This is then the appropriate definition of total momentum for a charged particle in the presence of a magnetic field.

We saw in Chapter 1 that the transition from classical to quantum mechanics is made by replacing the momentum p by the operator $-i\hbar\nabla$, with

$$p\psi = -i\hbar\nabla\psi$$ (8.14)

As p is defined in terms of the gradient operator, this implies that p is completely determined by the geometry of the wavefunction (a more rapidly varying wavefunction implies larger total momentum).

In isolated atoms, the electronic wavefunction is rigid, and unchanged to first order in an applied magnetic field, B. We will see below that the same must also be true for superconducting electrons. The total momentum p is therefore conserved in an applied magnetic field. The average electron velocity must be zero ($\langle v \rangle = 0$) when the applied vector potential $A = 0$, giving $p = 0$. The conservation of total momentum then requires $Mv + qA = 0$, so that for an electron of mass m_e and charge $-e$, we can write

$$v = \frac{eA}{m_e}$$ (8.15)

in an applied vector potential A. The resulting induced current density j is then given by

$$j = n(-e)v = -\frac{ne^2}{m_e} A$$ (8.16)

where n is the density of electrons per unit volume. By assuming that we can associate a rigid (macroscopic) wavefunction with the n_s superconducting electrons per unit volume in a superconductor we derive that

$$j = \frac{-n_s e^2}{m_e} A$$ (8.17)

Taking the curl of both sides we find that the magnetic field B and current density j are related in a superconductor by

$$\nabla \times j = -\frac{n_s e^2}{m_e} B \qquad (8.18)$$

This relation was deduced from the Meissner effect by Fritz and Heinz London in 1935, and is referred to as the London equation. We have shown here how the London equation follows from the assumption of a rigid wavefunction, and will show below how it leads to the Meissner effect.

8.6 The Meissner effect

To determine the variation of the magnetic field, B, within a superconductor, we can combine the London equation (eq. (8.18)) with Maxwell's steady-state equations for the magnetic field:

$$\nabla \times B = \mu_0 j \qquad (8.19a)$$

$$\nabla \cdot B = 0 \qquad (8.19b)$$

Taking the curl of the first of these Maxwell equations, we find

$$\nabla \times (\nabla \times B) = \mu_0 (\nabla \times j) \qquad (8.20)$$

We use the London equation to replace $\nabla \times j$ on the right-hand side of eq. (8.20), and use the vector identity $\nabla \times (\nabla \times B) = \nabla(\nabla \cdot B) - \nabla^2 B$ on the left-hand side. This gives

$$\nabla(\nabla \cdot B) - \nabla^2 B = -\frac{\mu_0 n_s e^2}{m_e} B \qquad (8.21)$$

Substituting the second Maxwell equation, eq. (8.19b), into eq. (8.21), we find a second order equation determining the behaviour of the magnetic field, B,

$$\nabla^2 B = \lambda_L^{-2} B \qquad (8.22)$$

where λ_L is referred to as the London penetration depth, and is given by

$$\lambda_L^2 = \frac{m_e}{\mu_0 n_s e^2} \qquad (8.23)$$

By taking the curl of the London equation, we can derive a similar expression for the current density,

$$\nabla^2 j = \lambda_L^{-2} j \qquad (8.24)$$

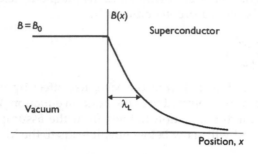

Figure 8.7 A magnetic field decays exponentially inside the surface of a superconductor, with the decay length, λ_L, given by the London equation.

The Meissner effect follows immediately from eq. (8.22), because $B = C$ is not a solution to eq. (8.22) unless the constant, $C = 0$.

The London equation predicts the exponential decay of a magnetic field B away from the surface of a superconductor. We can see this by considering a semi-infinite superconductor, defined in the half-space $x > 0$, and with a constant magnetic field, $B = B_0$ in the vacuum region ($x < 0$), with the field B_0 pointing along the z-direction (fig. 8.7). Because the field is independent of y and z, eq. (8.22) reduces to

$$\frac{\partial^2 B}{\partial x^2} = \frac{1}{\lambda_L^2} B \qquad (8.25)$$

which has the general solution,

$$B(x) = A \exp(x/\lambda_L) + B \exp(-x/\lambda_L) \qquad x > 0 \qquad (8.26)$$

Applying the boundary conditions that B remain finite as $x \to \infty$, and $B = B_0$ at $x = 0$, we then find that the magnetic field B decays exponentially into the semi-infinite superconductor as

$$B(x) = B_0 \, e^{-x/\lambda_L} \qquad (8.27)$$

The current density j also decays exponentially into the superconductor, as $j(x) = j_0 \exp(-x/\lambda_L)$, with a typical value of the London penetration depth, λ_L, being of order 300 Å.

To re-emphasise that a superconductor is more than just a perfect conductor, we note that in a perfect conductor the force F on an electron of mass m_e and charge $-e$ is given by

$$F = m_e \frac{dv}{dt} = -eE \qquad (8.28)$$

where E is the instantaneous electric field at any point. The current density $j = -n_s e v$ is therefore related to E by

$$\frac{\mathrm{d}j}{\mathrm{d}t} = \frac{n_s e^2}{m_e} E \tag{8.29}$$

and as $\nabla \times E = -\partial B/\partial t$, we find by taking the curl of both sides of eq. (8.29) that

$$\frac{\mathrm{d}}{\mathrm{d}t}(\nabla \times j) = -\frac{\mathrm{d}}{\mathrm{d}t}\left(\frac{n_s e^2}{m_e} B\right) \tag{8.30}$$

so that

$$\frac{\mathrm{d}}{\mathrm{d}t}\left(\nabla \times j + \frac{n_s e^2}{m_e} B\right) = 0 \tag{8.31}$$

A perfect conductor therefore requires that $(\nabla \times j + (n_s e^2/m_e)B)$ is *constant* with time. From the London equation, a superconductor requires that this constant is zero.

8.7 Application of thermodynamics

We saw earlier that the transition from the superconducting to normal state occurs at a critical field H_c for a Type I superconductor, and that H_c decreases with increasing temperature T (fig. 8.4(b)). Flux is always excluded from the superconductor no matter how we approach the superconducting state. Because the transition from the normal to the superconducting state is reversible, we can use thermodynamic analysis to investigate the superconducting transition. This allows us to deduce the energy difference between the normal and superconducting states, which turns out to be remarkably small. We can also determine the difference in entropy, or degree of disorder, between the two states. The results of the thermodynamic analysis then place severe constraints on the possible models of superconductivity.

The *Gibbs free energy* per unit volume, G, is the thermodynamic function which must be minimised to determine the equilibrium state of any substance at a fixed temperature T, pressure p and applied field H (see, e.g. Finn). It is defined for a magnetic material with magnetisation M in an applied field H by

$$G = U - TS + pV - \mu_0 H \cdot M \tag{8.32}$$

where U is the internal energy of the substance, and S the entropy, which is a measure of the degree to which the substance is disordered. (S increases as the disorder increases.) Work must be carried out to create a change in magnetisation dM in the external field H. The change in the internal energy per unit volume, dU, therefore depends on temperature T and applied field H as (Zemansky and Dittman (1981))

$$dU = T \, dS + \mu_0 H \cdot dM (-p \, dV) \tag{8.33}$$

Taking the differential of eq. (8.32) and ignoring the change in G or U due to any small changes in volume or pressure we find that the Gibbs free energy changes with applied field H and temperature T as

$$dG = dU - d(TS) - \mu_0 \, d(M \cdot H)$$
$$= -\mu_0 M \cdot dH - S \, dT \tag{8.34}$$

so that for a constant temperature T, the change in Gibbs free energy per unit volume, dG, is given by

$$dG = -\mu_0 M \cdot dH \tag{8.35}$$

We can use eq. (8.35) to determine the change in Gibbs free energy per unit volume of a superconductor, G_{sc}, as a function of applied magnetic field H

$$G_{sc}(H) - G_{sc}(0) = \int_0^H (-\mu_0 M) \, dH$$
$$= \int_0^H \mu_0 H dH \tag{8.36}$$

because $M/H = -1$ in a perfect diamagnet. We therefore conclude that the Gibbs free energy increases with applied field H in a superconductor as

$$G_{sc}(H) = G_{sc}(0) + \tfrac{1}{2}\mu_0 H^2 \tag{8.37}$$

At the critical field, H_c, the Gibbs free energy of the superconducting and normal states are equal

$$G_{sc}(H_c) = G_N(H_c) \tag{8.38}$$

while above H_c, $G_N < G_{sc}$, and the normal state is the equilibrium state.

Because the normal state is effectively non-magnetic ($M \sim 10^{-5}H \approx 0$), we have

$$G_N(0) \simeq G_N(H_c) \tag{8.39}$$

Combining equations (8.37), (8.38), and (8.39) the difference in Gibbs free energy between the normal and superconducting states at zero field is, therefore,

$$G_N(0) - G_{sc}(0) = \frac{1}{2}\mu_0 H_c^2 \qquad (8.40)$$

Applying this equation to the experimentally observed temperature dependence of H_c (fig. 8.4a) we can deduce several important features concerning the comparative entropy and energy of the superconducting and normal states.

The entropy, S, of a substance is defined in terms of its Gibbs free energy as (Finn (1993); Zemansky and Dittman (1981))

$$S = -\left(\frac{\partial G}{\partial T}\right)_{H,p} \qquad (8.41)$$

Applying this definition to eq. (8.40) we find that

$$S_N(0, T) - S_{sc}(0, T) = \frac{-\partial}{\partial T}\left(\frac{1}{2}\mu_0 H_c^2\right)$$

$$= -\mu_0 H_c \frac{\partial H_c}{\partial T} \qquad (8.42)$$

we have that $H_c = 0$ at the superconducting transition temperature, so that the difference in entropy, ΔS, equals zero at $T = T_c$. Similarly, as $T \rightarrow 0$, $\partial H_c/\partial T \rightarrow 0$, so that ΔS also equals zero at $T = 0$, that is, the superconducting and normal states are equally ordered at $T = 0$ and at $T = T_c$. We note, however, that at all intermediate temperatures ($0 < T < T_c$) $\partial H_c/\partial T < 0$, so that $\Delta S > 0$. The superconducting state is therefore more ordered than the normal state (fig. 8.8).

A major challenge in developing a microscopic theory of superconductivity was to account for the very small energy difference between the superconducting and normal states. From eq. (8.40), the energy difference per unit volume between the normal and superconducting states is given by

$$\Delta G = \frac{1}{2}\mu_0 H_c^2 = \frac{B_c^2}{2\mu_0} \qquad (8.43)$$

where B_c is the critical field in Tesla. We take aluminium as a typical Type I superconductor, for which $B_c = 0.0105\,\text{T}$ and in which there are approximately $N = 6 \times 10^{28}$ conduction electrons per cubic metre. The average energy difference per electron, ε_1, between the superconducting and normal states is given by $\varepsilon_1 = \Delta G/N \sim 10^{-8}\,\text{eV}$ per electron. This is considerably lower than thermal energies at the critical temperature, kT_c,

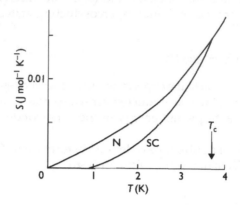

Figure 8.8 Variation of the entropy, S, with temperature in the normal (N) and superconducting (SC) states of a metal (after Keesom and van Laer 1938).

which are of order 10^{-4} eV, indicating it is most unlikely that all of the electrons gain energy in the superconducting transition.

We can instead apply an argument analogous to that used when estimating the paramagnetic susceptibility of a metal in Chapter 6, where we saw that electrons within an energy $\mu_B B$ of the Fermi energy gained energy of order $\mu_B B$ by flipping their spin direction. We presume here that the superconducting transition is due to those electrons within an energy ε_2 of the Fermi energy, E_F, and that each of these electrons gains energy of order ε_2 through the superconducting interaction. We estimate the number of such electrons per unit volume, n_{sc}, as

$$n_{sc} \approx g(E_F)\varepsilon_2 \tag{8.44}$$

where $g(E_F)$ is the density of states at the Fermi energy in the normal metal. It can be shown from eq. (5.15) that $g(E_F) = \frac{2}{3}(N/E_F)$ for a free-electron metal. Substituting this into eq. (8.44), we can estimate that

$$\frac{n_{sc}}{N} \sim \frac{\varepsilon_2}{E_F} \tag{8.45}$$

where we drop the factor of $\frac{2}{3}$, because we are just making an order of magnitude estimate. The total energy gained per unit volume by these n_{sc} superconducting electrons equals $N\varepsilon_1 (=\frac{1}{2}\mu_0 B^2)$, $n_{sc} \varepsilon_2 = N\varepsilon_1$, so that

$$\varepsilon_2 \sim \frac{E_F \varepsilon_1}{\varepsilon_2} \tag{8.46}$$

Re-arranging, we then find

$$\varepsilon_2 \sim \sqrt{\varepsilon_1 E_F} \tag{8.47}$$

The Fermi energy $E_F \sim 1-10\,\text{eV}$ in aluminium, so that $\varepsilon_2 \sim 10^{-4}\,\text{eV}$. This is comparable to the thermal energy at the superconducting transition temperature, $\varepsilon_2 \sim kT_c$.

Thermodynamic analysis, therefore, suggests that the superconducting transition is due to those electrons within an energy range kT_c of the Fermi energy, with each of these electrons gaining $\sim kT_c$ of energy in the superconducting state. Furthermore, the superconducting state contains a more ordered arrangement of electrons than that which is found in the normal state.

8.8 Cooper pairs and BCS theory

The major breakthrough to explain the phenomenon of superconductivity came in the 1950s, when Leon Cooper showed that it is possible for a net attraction between a pair of electrons in a metal to bind the electrons to each other, forming a so-called 'Cooper pair' (1956). This idea was built upon by Bardeen, Cooper, and Schrieffer (1957). The model which they developed, referred to as BCS theory, provides in the main a very satisfactory explanation of conventional superconductors, and also points towards some of the requirements of a model for high-T_c superconductors. We present here a qualitative description of BCS theory. The mathematical details of the model are more advanced than we wish to consider, but are well described in a number of other texts (e.g. Ibach and Lüth 1995; Madelung 1978; Tinkham 1996).

An electron passing through a crystal lattice can cause a transient distortion of the positive lattice ions as it passes them (fig. 8.9). This vibrational distortion of the lattice can then attract another electron. The distortion-induced attraction is opposed by the short-range Coulomb repulsion between the two negative electrons. Broadly speaking, conventional superconductivity occurs in those materials for which the transient, lattice-mediated attraction is stronger than the Coulomb repulsion. Just as the quantum unit of light is referred to as a photon, so the quantum unit of vibration is referred to as a phonon, with the energy of a single phonon with frequency v equal to hv. Typical maximum phonon energies are of order $20-30\,\text{meV}$. It is reasonable to expect that only electrons within an energy range hv of the Fermi energy, E_F, are likely to be influenced by transient lattice distortions.

The establishment of the superconducting state by cooling to below the critical temperature, T_c, is a cooperative effect. The ground state, lowest-energy, configuration is achieved by all the superconducting electrons

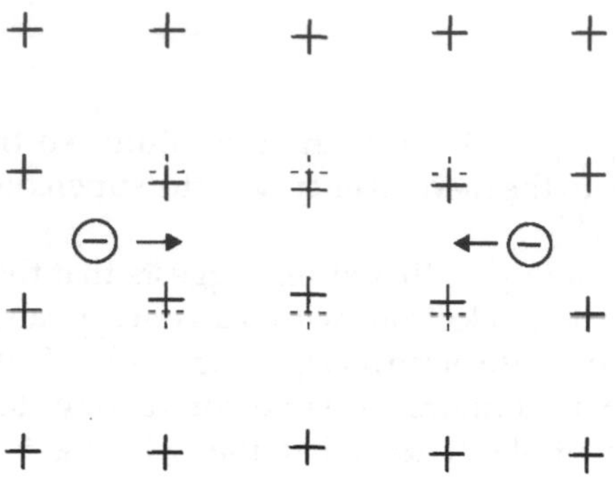

Figure 8.9 An electron passing through a crystal can cause a transient distortion of the neighbouring positive ions. This distortion of the lattice (= creation of a phonon) can attract another electron, thereby giving a net attraction between the two electrons.

forming identical Cooper pairs. The two electrons have equal and opposite momentum and spin, so that each Cooper pair has no net momentum or spin.

Because of the attraction between the electrons, a finite amount of energy is needed to break a Cooper pair. BCS theory, therefore, predicts an energy gap (whose magnitude is defined as 2Δ) between the highest filled Cooper pair and the next available (single electron) state above the energy gap. Figure 8.10 shows schematically the density of states at zero temperature of a normal metal and of a superconductor. It should be noted, however, that the energy gap in the superconductor has both different origins and a different meaning to that of a semiconductor in earlier chapters. The states immediately below the superconductor energy gap are *not* single-particle states, as in the semiconductor. Rather they are Cooper pair states.

Figure 8.11 shows that the superconducting energy gap, 2Δ, is temperature dependent. This arises because of the cooperative nature of superconductivity: adding energy to the metal breaks some Cooper pairs, thus making it easier to break further Cooper pairs, with the superconducting electron density, n_s varying with temperature approximately as

$$n_s(T) = n_s(0)\left[1 - \left(\frac{T}{T_c}\right)^4\right] \tag{8.48}$$

This is directly analogous to the way in which the spontaneous magnetisation varies with temperature in a ferrromagnet: a large

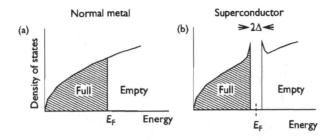

Figure 8.10 Schematic diagram of the density of states at $T = 0$ in (a) a normal metal, and (b) a superconductor. In both cases, all states are filled up to the Fermi energy, E_F, and are empty above E_F. There is an energy gap in the superconductor, defined to be of magnitude $2\Delta(0)$, between the highest filled Cooper pair state and the lowest empty single particle state.

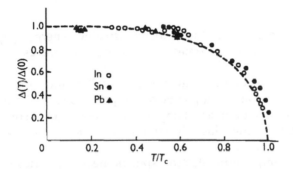

Figure 8.11 The temperature dependence of the superconducting energy gap, $\Delta(T)/\Delta(0)$, as predicted by BCS theory (dashed line) and as measured experimentally (after Giaever and Megerle 1961).

magnetisation is maintained until close to the Curie temperature, but the magnetisation then decreases rapidly approaching the Curie temperature, as the net number of spins contributing to the cooperative interaction decreases. So too, the density of Cooper pairs drops rapidly approaching the superconducting transition temperature.

We saw earlier that the London penetration depth, λ_L, varies with the superconducting electron density, n_s, as $\lambda_L \sim n_s^{-1/2}$ (eq. (8.23)). Combining this with eq. (8.48) BCS theory then predicts that λ_L should vary with temperature approximately as $[1 - (T/T_c)^4]^{-1/2}$. This relation is in good agreement with experimental observation for many superconductors.

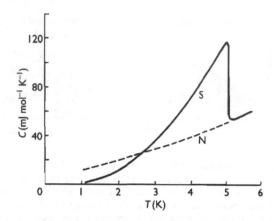

Figure 8.12 The variation in the heat capacity, c_V, of a superconducting (S) and normal metal (N), at and below the superconducting tranisition temperature, T_c (after Corak *et al.* 1956).

Further evidence for the superconducting energy gap, and its temperature dependence, comes from a variety of sources.

1   A normal metal absorbs microwaves and far infra-red radiation, by exciting electrons from just below to just above the Fermi energy. In a superconductor at zero temperature, there is an absorption edge at $h\nu = 2\Delta(0)$ below which the superconductor is perfectly reflecting to incident photons.

2   The low-temperature electron specific heat, c_V, varies exponentially with temperature, as $c_V \sim \exp(-\Delta/kT)$ in a superconductor, compared to a linear variation, $c_V \sim T$, in a normal metal (fig. 8.12). The difference arises because energy can only be added to the electrons in the superconductor by exciting electrons across the energy gap, breaking Cooper pairs to create single electron states above the Fermi energy, E_F.

3   Consider two normal metals separated at low temperature by a very thin insulating layer (fig. 8.13a(i)). For a sufficiently thin insulator, the two metals will share the same Fermi energy, E_F (fig.8.13a(i)). If a voltage, V, is now applied across the structure, most of the potential drop will occur across the insulating layer (fig.8.13a(ii)), giving rise to a current due to electrons tunnelling from one metal to the other. The current is predicted, and observed, to increase linearly with applied field, V, due to the linear shift in the the relative positions of the Fermi energies (fig. 8.13a(iii)). By contrast, when two superconductors are separated by a thin insulator (fig. 8.13b(i)), little current is observed at a very low applied voltage V (fig. 8.13b(iii)), because electrons cannot

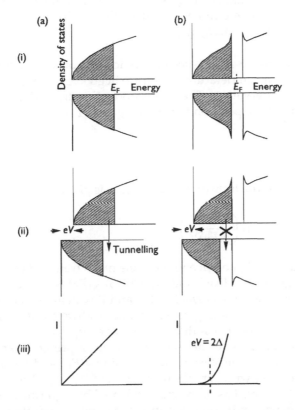

Figure 8.13 Comparison of I–V characteristics of (a) two normal metals, and (b) two superconductors separated by a thin insulating layer. (a)(i) When $V = 0$, the two metals share a common Fermi energy, E_F, and there is no net flow of carriers. (a)(ii) When $V \neq 0$, the voltage is dropped mainly across the resistive insulating layer, and carriers can tunnel from filled (shaded) states in one metal to empty states in the second metal, leading (a)(iii) to a linear increase in current, I, with applied voltage, V. (b)(i) For $V = 0$, the two superconductors also share a common E_F. (b)(ii) For small applied voltage, ($eV < 2\Delta$), the voltage drop across the insulating layer is insufficient to align filled (Cooper pair) states in one superconductor with empty (single particle) states in the second, so that (b)(iii) the current only starts to increase significantly when $eV > 2\Delta$.

tunnel from the filled states in the first superconductor through to the energy gap of the second. The tunnelling current is observed to switch on sharply at an applied voltage, V such that $eV = 2\Delta$, at which point electrons can tunnel from the filled (Cooper pair) states in the first superconductor through to the empty (single-particle) states in the second superconductor (fig. 8.13b(iii)). Such tunnelling experiments can be used to determine Δ, and further verify the existence of the superconducting energy gap.

BCS theory makes several specific predictions concerning the superconducting transition temperature, which are in good general agreement with experimental observation. The superconducting transition temperature, T_c, is predicted to depend on the energy gap at $T = 0\,\mathrm{K}$ as (see e.g. Ibach and Lüth 1995)

$$2\Delta(0) = 3.53\,kT_c \tag{8.49}$$

More specifically, the transition temperature is given by

$$kT_c = 1.13\,hv\,\exp\left(\frac{-1}{g(E_F)V}\right) \tag{8.50}$$

where hv is a typical phonon energy in the metal (referred to as the Debye phonon energy, see e.g. Ibach and Lüth 1995), $g(E_F)$ is the density of states at the Fermi energy in the normal metal, and V is an interaction parameter. $g(E_F)V$ is called the coupling constant, and is always less than 1, so that kT_c is then always much less than the Debye energy, hv.

Equation (8.50) supports the qualitative analysis presented above, which noted that T_c should increase with increasing $g(E_F)$ and hv. Since vibrational energies vary with particle mass, M, as $M^{-1/2}$, eq. (8.50) then implies that $T_c \propto M^{-1/2}$. This has been well confirmed by experimental measurements of the superconducting transition temperature for different isotopes of elements such as tin and lead. It is referred to as the isotope effect, and is one of the key pieces of evidence supporting BCS theory.

8.9   BCS coherence length

The average distance, ξ_0, between the electrons in a Cooper pair at $T = 0$ is of order

$$\xi_0 = \frac{\hbar v_F}{\pi \Delta(0)} \tag{8.51}$$

where v_F is the electron velocity at the Fermi energy, and ξ_0 is known as the BCS coherence length. This definition of the coherence length is consistent with a dimensional analysis. If the electrons each gain energy $\Delta(0)$ through their interaction, then they must be coherent with each other at least over a timescale $\tau \sim \hbar/\Delta(0)$ (eq. (1.16)). In this time, the electrons at the Fermi energy can travel a distance $v_F\tau$, comparable to the coherence length. The formation of alloys, or introduction of impurities reduces the magnitude of ξ_0.

The relative magnitude of ξ_0 and the London penetration depth, λ_L, plays the major role in determining whether a particular superconductor will be Type I or Type II. In general, $\xi_0 > \lambda_L(0)$ in Type I superconductors, while $\xi_0 < \lambda_L(0)$ in Type II superconductors. This occurs because the concentration of Cooper pairs changes gradually, over a coherence length $\sim\xi_0$. When $\xi_0 < \lambda_L$, it is possible for flux vortices to penetrate through a superconductor, as the superconducting electron density can adjust over the transition region of width λ_L between the normal and superconducting states inside and outside the vortex, respectively. By contrast, this is not possible when $\xi_0 > \lambda_L$, the condition for a Type I superconductor.

8.10 Persistent current and the superconducting wavefunction

We have seen above that the number of Cooper pairs decreases with increasing temperature, as electrons are thermally excited across the super-conducting energy gap. It is therefore surprising, given this constant breaking of Cooper pairs, as to why a persistent current is found in a super-conductor. The answer is that Cooper pairs are also continuously being formed, as single electrons bind to each other through the superconduct-ing interaction. Because of the cooperative nature of this interaction, energy is only gained if these new Cooper pairs are in the same state as the existing pairs. This dynamic equilibrium allows for the continued existence of the superconducting current.

Because all the superconducting electrons are effectively in the same state, we can define a macroscopic superconducting wavefunction, or order parameter, $\psi(r)$, related to the superconducting electron density, $n_s(r)$ by

$$|\psi(r)|^2 = \tfrac{1}{2}n_s(r) \tag{8.52}$$

where $n_s/2$ is the Cooper pair density.

8.11 Flux quantisation

This macroscopic order parameter has properties similar to a quantum mechanical wavefunction. If the order parameter varies locally as

$$\psi(r) = \psi_0\, e^{i q \cdot r} \tag{8.53}$$

then we can associate a momentum $2m_e v = \hbar q$ with each Cooper pair, with the supercurrent density, $j_s(r)$ given by

$$
\begin{aligned}
j_s(r) &= \frac{n_s}{2}(-2e)\frac{\hbar q}{2m_e} \\
&= -|\psi(r)|^2 \frac{e}{m_e}\hbar q
\end{aligned}
\tag{8.54}
$$

We must modify this definition of current density in the presence of an applied magnetic field, B. We saw in Section 8.5 that the electromagnetic momentum p for a particle (Cooper pair) of mass $2m_e$ and charge $-2e$ is given by eq. (8.6) as

$$
p = 2m_e v - 2eA \tag{8.55}
$$

where A is the magnetic vector potential, and $B = \nabla \times A$. The kinetic momentum is therefore given by

$$
2m_e v = p + 2eA \tag{8.56}
$$

and the velocity v of a Cooper pair by

$$
v = \frac{p}{2m_e} + \left(\frac{e}{m_e}\right)A \tag{8.57}
$$

To calculate the velocity v in a quantum mechanical analysis, we replace the total momentum p by the momentum operator $-i\hbar\nabla$. If we assume that this is also true for the superconducting order parameter, we might deduce from eq. (8.54) that the superconducting current density $j(r)$ in the presence of a magnetic field B is given by

$$
\begin{aligned}
j(r) &= \psi^*(r)(-2ev)\psi(r) \\
&= -\frac{e}{m_e}\psi^*(r)(-i\hbar\nabla + 2eA)\psi(r)
\end{aligned}
\tag{8.58}
$$

This analysis is only partly correct. It is possible using eq. (8.58) to find sensible wavefunctions (e.g. $\psi(r) = \sin(k \cdot r)$) for which the calculated current density is imaginary. A more complete analysis (see e.g. Hook and Hall 1991) shows that in order for the local current density to be real (and therefore a measurable quantity), we must define $j(r)$ as half the sum of eq. (8.58) and its complex conjugate:

$$
j(r) = \frac{i\hbar e}{2m_e}(\psi^*(r)\nabla\psi(r) - \psi(r)\nabla\psi^*(r)) - \frac{2e^2}{m_e}A\psi^*(r)\psi(r) \tag{8.59}
$$

We can choose to write the superconducting order parameter at each point as the product of a real number times a phase factor,

$$
\psi(r) = |\psi(r)|e^{i\theta(r)} \tag{8.60}
$$

When we do so, eq. (8.59) can be simplified to give

$$j(r) = -\left(\frac{e}{m_e}\right)|\psi(r)|^2(\hbar\nabla\theta + 2eA) \tag{8.61}$$

This equation is consistent with our earlier analysis. It reduces to eq. (8.54) if we let $\theta = k \cdot r$, set $A = 0$, and replace $|\psi(r)|^2$ by $n_s/2$, the Cooper pair density. Also, if we assume that $|\psi(r)|^2$, and hence n_s is constant, and take the curl of eq. (8.61) we recover the London equation.

One remarkable consequence of eq. (8.61) is that the magnetic flux penetrating a superconducting ring is quantised: we can directly measure a quantum effect in a macroscopic sample. Consider a superconducting ring, as in fig. (8.14), which carries a persistent supercurrent. From the London equation, we know that the superconducting current only flows on the outer surface of the ring, so that $j(r) = 0$ on the dotted path in fig. (8.14). Because the superconducting electron density, and hence $\psi(r)$, is finite throughout the sample, this then requires from eq. (8.61) that

$$\hbar\nabla\theta = -2eA \tag{8.62}$$

Integrating around the closed dotted path, C, within the superconductor then gives

$$\oint_C \hbar\nabla\theta \cdot dl = -\oint_C 2eA \cdot dl \tag{8.63}$$

Because the order parameter, $\psi(r)$, can only have one value at each point within the superconductor the phase θ must change by an integer multiple of 2π going round the closed loop. The left-hand side of eq. (8.63)

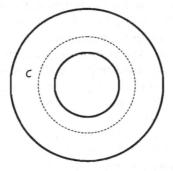

Figure 8.14 Because superconducting current only flows within a distance of order λ of the surface of the ring shown, the current density $j(r) = 0$ everywhere on the dotted loop C within the body of the ring. We can use this to prove that the magnetic flux Φ is quantised through the centre of the loop.

then equals $2\pi\hbar n = hn$, where n is an integer, while the right-hand side equals $-2e\Phi$, where Φ is the magnetic flux linking the loop (eq. (8.9)). Re-arranging eq. (8.63) we find that the magnetic flux Φ linking a closed superconducting ring is quantised, with

$$\Phi = \frac{hn}{2e} \qquad (8.64)$$

where the factor of $2e$ arises because the electrons are bound together in Cooper pairs. Flux quantisation has been well confirmed experimentally for conventional superconductors.

8.12 Josephson tunnelling

We saw in Section 8.8 that when two superconductors are separated by an insulating layer, then electrons can tunnel through the layer, allowing current to flow for a sufficiently large applied voltage. Even more striking effects are observed if the insulating layer is very thin, so that the super-conducting order parameters, $\psi(r)$ from both sides of the layer become weakly coupled. This effect was first predicted in a short paper by Brian Josephson, based on a theoretical analysis he carried out while still a student at Cambridge (Josephson 1962), and for which he was awarded the Nobel prize in 1973.

If we place an insulating layer next to a superconductor, the superconducting order parameter, $\psi(r)$ will decay exponentially into the insulating layer, analogous to the exponential decay of a bound state wavefunction outside a square quantum well (Chapter 1). If two superconducting regions are separated by a thick insulating layer there will be no overlap between their order parameters in the insulating layer; the two superconducting regions will be decoupled, and act independently of each other. However, for a sufficiently thin insulating layer (~ 10Å) the superconducting electron density will never drop fully to zero in the insulating layer, and so the two regions will be weakly coupled, as illustrated in fig. 8.15(a), where the weak link is formed by oxidising a small cross-section of what would otherwise be a superconducting ring. Because both sides of the link are at the same temperature, the superconducting electron density, and hence the magnitude of the superconducting order parameter should be equal on each side of the link, with magnitude say ψ_0. We define the insulating layer in the region $-b/2 \leq x \leq b/2$, and allow for a difference in the phase, θ, of the order parameter on either side of the link, with $\theta = \theta_L$ on the left-hand side and $\theta = \theta_R$ on the right-hand side. The superconducting order parameter then varies in the insulating layer as (fig. 8.15(b))

$$\psi(r) = \psi_0(e^{i\theta_L - \kappa(x+d/2)} + e^{i\theta_R + \kappa(x-d/2)}) \qquad (8.65)$$

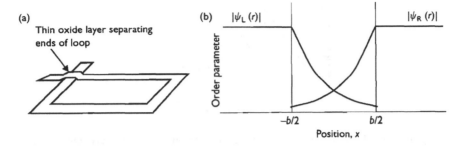

(a)
Thin oxide layer separating
ends of loop

(b) $|\psi_L(r)|$ $|\psi_R(r)|$

Order parameter

$-b/2$ $b/2$

Position, x

Figure 8.15 (a) A weak link is formed in a superconducting loop by interrupting the loop
at one point by a very thin oxide layer ($\sim$10 Å across). The superconducting
order parameter decays exponentially from either side into the insulating layer,
varying as shown in (b).

The order parameter $\psi(r)$ of eq. (8.65) allows a superconducting current to
flow even in the absence of an applied voltage or external magnetic field
($B = 0$). We substitute eq. (8.65) into eq. (8.59), and set $A = 0$ to derive the
DC Josephson effect, whereby

$$j(r) = j_0 \sin(\theta_L - \theta_R) \qquad (8.66a)$$

with

$$j_0 = \frac{e\hbar n_s \kappa \, e^{-\kappa d}}{m_e} \qquad (8.66b)$$

The difference in the order parameter phase factor on either side of a weak
link is therefore directly related to the supercurrent density in the weak
link. Equation (8.66) indicates that a DC supercurrent can flow even in the
absence of an applied voltage or external magnetic field.

The DC Josephson effect provides a very accurate means of measuring magnetic flux, using what is known as a SQUID magnetometer, that is,
a *Superconducting QUantum Interference Device*. Consider the superconducting ring with a weak link illustrated in fig. (8.15a). The total magnetic
flux Φ linking this ring must at all times equal $n\Phi_0$, where n is an integer, and Φ_0 is the superconducting unit of magnetic flux, $h/2e$ (eq. (8.64)).
Let the ring sit in an external magnetic field, B, chosen such that the flux
through the ring equals $n\Phi_0$. If the external field is now increased, a superconducting current will be induced in the ring, whose magnitude varies
linearly with the change in B in order to maintain the flux $n\Phi_0$. However, because the superconducting carrier density, n_s, is lower in the weak
link, the superconducting current can be more easily destroyed in this
region. As the current is destroyed, the flux through the loop can change,
increasing to $(n+1)\Phi_0$, and restoring the superconducting link. The current

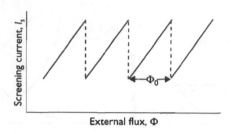

Figure 8.16 Schematic illustration of variation of screening current with external magnetic flux linking a superconducting loop with one weak (Josephson)
link.

circulating in the ring, therefore, varies with applied field as indicated
schematically in fig. 8.16. A more detailed analysis of the principles of the
SQUID magnetometer is provided in several text books (e.g. Myers 1997;
Hook and Hall 1991; Tanner 1995). By appropriate design of the magnetometer, it is possible to measure magnetic field to very high sensitivity, of
order 10^{-11} G or less. Such devices are now available commercially. Mag-
netometers can also be designed to measure very small changes in field
gradient, achieved by having an external double coil, where the two coils
are wound in opposite directions. A constant field then gives rise to equal
and opposite flux in each half of the coil, with the double coil then only
sensitive to changes in the field, that is, the field gradient (Tanner).

8.13   AC Josephson effect

We write the superconducting order parameter as $\psi(r)$ because of its
similarity to the quantum mechanical wavefunctions found by solving
Schrödinger's equation. We saw in Chapter 1 that if $\phi_n(r)$ is a solution of
the steady-state Schrödinger equation with energy E_n then we can describe
the time dependence of this state by

$$\Phi_n(r,t) = \phi_n(r)\exp(iE_n t/\hbar) \tag{8.67}$$

Likewise the superconducting order parameter of a state with energy E_i
varies with time as

$$\Psi(r,t) = |\psi(r)|\exp(i\theta(t)) \tag{8.68}$$

where the phase factor $\theta(t)$ is given by

$$\theta(t) = \theta_0 + \frac{E_i t}{\hbar} \tag{8.69}$$

When we apply a DC voltage, V, to a superconducting circuit containing a
Josephson junction, all of the voltage will be dropped across the weak link

in the circuit, namely the Josephson junction, giving an energy difference $\Delta E = 2eV$ across the junction, so that

$$\theta_{\mathrm{L}}(t) - \theta_{\mathrm{R}}(t) = \theta_0 + \frac{\Delta E t}{\hbar} \tag{8.70}$$

Substituting this into eq. (8.66a) we therefore deduce that a DC voltage, V, across a Josephson junction leads to an AC current flow,

$$j(\mathbf{r}, t) = j_0 \sin(\theta_0 + \frac{2eVt}{\hbar}) \tag{8.71}$$

with angular frequency $\omega = 2eV/\hbar$ and frequency $\nu = 2eV/h$. Measuring ν for a given applied voltage has allowed the determination of the ratio of the fundamental constants e and h to 1 part in 10^7, and is now also used as a means of measuring and calibrating voltage standards. The AC current also leads to the emission of electromagnetic radiation of frequency ν. (Energy is dissipated across the junction through the emission of a photon of energy $h\nu = 2eV$ each time a Cooper pair crosses the junction.) Although the power radiated by a single junction is very low, of order 10^{-10} W, Josephson junctions do nevertheless find some application as microwave power sources: an applied voltage of order 10^{-4} V leads to microwave emission about 50 GHz (wavelength, $\lambda \sim 6$ mm), with the emission wavelength also being tunable through variation of the applied voltage.

8.14 High-temperature superconductivity

The lure of high-T_{c} superconductors is partly psychological. These materials become virtually perfect conductors when plunged into liquid nitrogen at 77 K and, before one's very eyes, become capable of levitating a magnet. The discovery by Bednorz and Müller in late 1986 that the ceramic material, lanthanum barium copper oxide, lost all electrical resistance when cooled to 35 K gained them the Nobel prize in Physics within a year and unleashed an unparalleled explosion of research activity. Within eighteen months a wide range of further material combinations had been tested, leading to the discovery of compounds such as $Tl_2Ba_2Ca_2Cu_3O_{10}$ with a superconducting transition temperature as high as 125 K. Surprisingly, all of the high-temperature superconductors discovered to date share a common crystal structure: they all contain lightly doped copper-oxide layers, with other metal atoms sitting between these layers. Extensive research to find high-T_{c} superconductivity in other families of materials has been singularly unsuccessful. The cuprate family of materials continues, therefore, to be of immense theoretical and experimental interest. Even a decade and a half after its discovery, the mechanism underpinning

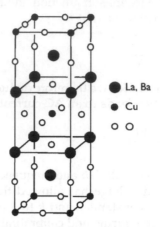

Figure 8.17 The crystal structure of $La_{1-x}Ba_xCuO_4$, an archetypal high-T_c super-conductor. Each copper atom is bonded to four neighbouring oxygen atoms within the plane. The copper oxide planes (the bottom, middle, and top layers in the structure shown) are separated from each other by layers containing La, Ba, and further oxygen atoms. Some high-T_c super-conductors with more complex crystal structures have two or three neighbouring copper oxide layers, with each group of copper oxide layers again separated from the next group of copper oxide layers by other metal oxide layers. (From H. P. Myers (1997) *Introductory Solid State Physics*, 2nd edn.)

high-T_c superconductivity has still not been resolved, although a wide range of measurements have established some of the key experimental features of these materials.

Figure 8.17 shows the crystal structure of $La_{2-x}Ba_xCuO_4$, whose structure and properties are typical of all high-temperature superconductors. Each copper atom lies in a 2D layer, bonded to four neighbouring oxygen atoms within the layer. The copper layers are then separated from each other by layers of lanthanum, barium, and oxygen atoms.

The electronic properties of $La_{2-x}Ba_xCuO_4$ vary dramatically as the alloy composition, x, is varied. When $x = 0$, La_2CuO_4 is an antiferromagnetic insulator, with a Néel temperature, $T_N = 240\,K$. The electronic structure can be described in terms of three electrons transferring from each lanthanum ion to the oxygen ions, with two electrons transferring from each of the copper ions, to give $(La^{3+})_2Cu^{2+}(O^{2-})_4$. The only ion which then possesses an incomplete shell of electrons is the Cu^{2+} ion, which has nine 3d electrons, just one electron short of a filled 3d shell. For further discussion, it is considerably more convenient to describe the Cu^{2+} ion as having one 3d hole state in an otherwise filled 3d band. This 3d hole state has a magnetic moment. Each hole is localised on a single copper ion, with the exchange interaction then leading to the formation of the anti-ferromagnetic insulator.

When lanthanum is replaced by barium, the charge balance is disturbed and an electron-deficient structure is formed. Each barium effectively donates an extra hole to the structure. It is thought that these extra holes are associated primarily with the oxygen atoms in the copper oxide layers. The holes are mobile, predominantly within the plane, so that as the barium concentration increases, the alloy becomes metallic and at the same time the Néel temperature decreases until eventually the material ceases to be an anti-ferromagnet (for $x \sim 0.05$). When a little more barium is added, the material becomes superconducting at low temperatures. The superconducting transition temperature, T_c, is maximised for $x \sim 0.2$, when there is about one extra hole for every five copper atoms. As x increases further, T_c starts to decrease, and for sufficiently high x (~ 0.3), the material no longer displays superconductivity.

From the above, it is clear that a model to explain high-T_c superconductors must, probably, be radically different from the BCS theory applicable to conventional superconductors. Early measurements of the flux quantum confirmed that the superconducting charge carriers have charge $2e$, so that hole pairing must occur. Nuclear magnetic resonance Knight shift measurements also provided evidence for anti-parallel alignment of the hole spins, as in conventional superconductors. More recently however it has been established that the spatial symmetry of the superconducting order parameter is markedly different in high-T_c superconductors compared to the conventional case. In a conventional superconductor, each Cooper pair is approximately spherically symmetric: there is equal probability of finding the second electron along any direction with respect to the first. By analogy with atomic wavefunctions, we therefore say that the order parameter has s-like symmetry. By contrast, a recent series of elegant experiments have shown the order parameter to have a markedly different, d-like, symmetry in high-T_c superconductors, as illustrated in fig. 8.18. The high-T_c order parameter has similar symmetry to a $d_{x^2-y^2}$ atomic orbital: each Cooper pair lies predominantly within a single CuO_2 plane; there is greatest probability of the two holes in the pair being oriented along the crystal axes, and zero probability of their lying along the (110) direction with respect to each other. Any model of high-T_c superconductivity must account for this d-like symmetry of the order parameter.

One testable property of d-wave symmetry is that the Cooper pairs are more weakly bound in some directions than in others, and so the superconducting energy gap should be angular dependent, going to zero along specific directions. This modifies the microwave absorption characteristics, and also introduces a quadratic term to the low-temperature specific heat. Initial measurements confirmed the angular dependence of the gap,

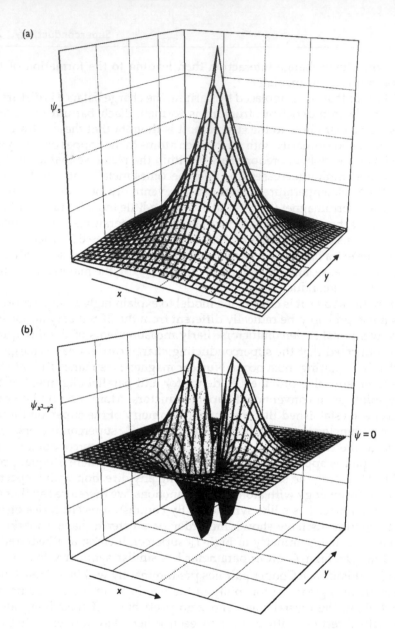

Figure 8.18 (a) s-wave symmetry of the superconducting order parameter in a conventional superconductor, where there is equal probability of finding the second electron in a Cooper pair along any direction with respect to the first. (b) Several experiments show that the order parameter has $d_{x^2-y^2}$ symmetry in high-T_c superconductors, meaning that the order parameter changes sign for a rotation of $\pi/2$ within the x–y plane, and that there is greatest probability of the two holes within the Cooper pair being oriented along the x- or y-axis with respect to each other.

but were not on their own sufficient to confirm d-wave symmetry, as they could also have been consistent with a modified s-like state.

The key feature of the d-wave order parameter is that its phase varies with direction, being of opposite sign along the x- and y-axes in fig. 8.18. The first tests for d-wave symmetry, probing the angular dependence of the energy gap, were insensitive to this phase variation, and so their results were suggestive but not conclusive. We saw earlier in Sections 8.11 and 8.12 that the magnetic flux through a closed loop depends on the total change in phase around the loop, and equals $n\Phi_0$ for a conventional superconductor, where $\Phi_0 = h/2e$. The magnetic flux linking a closed loop turns out to be a very useful probe of the order parameter symmetry, and has provided the clearest evidence so far for unconventional d-wave symmetry. Consider a superconducting circuit formed by linking an s-wave superconducting element with a d-wave element, as shown in fig. 8.19. The superconducting order parameter must vary continuously round this circuit. With zero

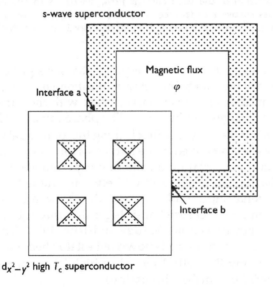

s-wave superconductor

Magnetic flux

φ

Interface a

Interface b

$d_{x^2-y^2}$ high T_c superconductor

Figure 8.19 Geometry for a superconducting circuit with two weak (Josephson) links between a $d_{x^2-y^2}$ high-T_c superconductor and a conventional s-wave superconductor. With zero external field and no current flow, the superconducting phase, θ, is constant in the s-wave superconductor (illustrated here as $\theta = 0$), but changes by π between the a and b faces of the high-T_c superconductor. The phase discontinuity of π indicated at the b interface is inconsistent with the general assumption that phase changes continuously. For continuous variation of phase, we therefore require a current flow, and conclude that the magnetic flux linking such a loop must equal $(n + \frac{1}{2})\Phi_0$. (from James Annett, *Contemporary Physics* **36** 433 (1995) © Taylor & Francis.)

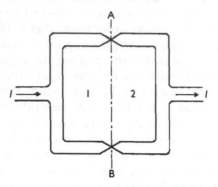

Figure 8.20 A loop in a superconducting circuit, with two Josephson junctions (labelled as A and B) in parallel. Interference between the superconducting order parameter on each side of the loop can lead to enhancement or cancellation of the total current flow, analogous to the constructive and destructive interference observed when light passes through a double slit. (From H. P. Myers (1997) *Introductory Solid State Physics*, 2nd edn.)

external field and no current flow, the phase changes sign between the a and b faces of the d-wave superconductor, while it remains constant within the s-wave part of the loop. This is inconsistent with the earlier assumption that the phase changes continuously. The phase can only change in a continuous manner if the magnetic flux linking the loop equals $(n + \frac{1}{2})\Phi_0$. This behaviour was first predicted in the late 1970s for a loop such as that in fig. 8.20. Wollman *et al.* (1993) found the first experimental evidence to support this behaviour in a loop formed between a high-T_c superconductor and a conventional superconductor. Shortly afterwards, Tsuei *et al.* (1994) were able to directly measure half-integer flux through a slightly different loop arrangement (see also Kirtley and Tsuei 1996). The experimental evidence for $d_{x^2-y^2}$ symmetry is now well established and must be taken into account in any theoretical model of high-T_c superconductivity (see e.g. Annett 1995 for a further discussion).

Several theoretical approaches are being taken to develop a model to account for the observed characteristics of high-T_c superconductors (Orenstein and Millis 2000). One promising approach is based on the idea of a doped resonant valence bond (RVB) state. The RVB state was first proposed by Philip Anderson to describe a lattice of antiferromagnetically coupled spins where the quantum fluctuations are so strong that long-range magnetic order is suppressed. The system resonates between states in which different pairs of spins form singlet states that have zero spin and hence no fixed spin direction in space. Many questions remain unanswered, however, such as how does a doped RVB state behave? And why does doping stabilise a RVB state when the undoped state prefers an ordered magnetic state? A comprehensive theory is currently lacking, but there are encouraging signs.

The RVB state is just one of many theoretical approaches to high-T_c superconductivity. Other competing theories include those based on fluctuating stripes. In certain cuprates at low temperature the doped holes are observed to localise along parallel lines, called stripes, in the copper-oxide planes (see e.g. Tranquada 1999). Another theory seeks to exploit the strong coupling between electrons and phonons in oxide materials, proposing the formation of entitities called bipolarons, giving a mechanism related but not identical to the pairing interaction in conventional superconductors, effectively an extreme case of the conventional BCS attraction (Alexandrov and Mott 1994). Finally we mention the antiferromagnetic spin fluctuation exchange model (Fisk and Pines 1998), in which spin waves replace phonons as the mediators of the attraction in a BCS-type model.

High-T_c superconductivity is a very fitting topic on which to end but not complete a textbook such as this, as at the time of writing no final theory exists to explain the underlying superconducting mechanism. We can nevertheless ask what the final theory should predict. First, it should describe the transition between the antiferromagnetic and superconducting phases as the doping is varied. Second, it should reveal the specific conditions in the cuprates that lead to this very special behaviour. From this should follow some suggestions for other materials that would show similar behaviour. While it may not be possible to predict T_c accurately – because, for instance, of a lack of precise input parameters – the final theory should give the correct order of magnitude for T_c and explain the trends that are observed in the cuprates. These challenges, combined with new experimental results, are likely to keep theorists busy for years to come, but hopefully not another decade and a half!

References

Alexandrov, A. S. and N. F. Mott (1994) *Reports on Progress in Physics* **57** 1197.
Annett, J. (1995) *Contemporary Physics* **36** 423.
Bardeen, J., L. N. Cooper and J. R. Schrieffer (1957) *Phys. Rev.* **108** 1175.
Bednorz, J. G. and K. A. Müller (1986) *Z. Physik B* **64** 189.
Bednorz, J. G. and K. A. Müller (1988) *Rev. Mod. Phys.* **60** 585.
Cooper, L. N. (1956) *Phys. Rev.* **104** 1189.
Corak, W. S., B. B., Goodman, C. B. Satterthwaite and A. Wexler (1956) *Phys. Rev.* **102** 656.
File, J. and R. G. Mills (1963) *Phys. Rev. Lett.* **10** 93.
Finn, C. B. P. (1993) *Thermal Physics* (Chapman and Hall), 2nd edn, London.
Fisk, Z. and D. Pines (1998) *Nature* **394**, 22.
Giaever, I. and K. Megerle (1961) *Phys. Rev.* **122** 1101.
Hook, J. R. and H. E. Hall (1991) *Solid State Physics*, 2nd edn, Wiley, Chichester.

Ibach, H. and H. Lüth (1995) *Solid State Physics* Springer, New York.
Josephson, B. D. *Phys. Lett.* **1** 251.
Keesom, W. H. and P. H. van Laer (1938) *Physica* **5** 193.
Kirtley, J. R. and C. C. Tsuei (August 1996) *Scientific American*, p. 50.
Madelung, O. (1978) *Introduction to Solid-State Theory*, Springer, New York.
Myers, H. P. (1997) *Introductory Solid State Physics*, 2nd edn, Taylor and Francis, London.
Nagamatsu, J. N. Nakagawa, T. Muranaka, Y. Zenitani and J. Akimitsu (2001) *Nature* **410** 63.
Orenstein, J. and A.J. Millis (2000) *Science* **288** 468.
Tanner, B. K. (1995) *Introduction to the Physics of Electrons in Solids*, Cambridge University Press.
Tinkham, M. (1996) *Introduction to Superconductivity*, McGraw-Hill, New York.
Tranquada, J. (1999) *Physics World* **12**(11) 19.
Tsuei, C. C., J. R. Kirtley, C. C. Chi, L. S. Yu-Jahnes, A. Gupta, T. Shaw, J. Z. Sun and M. B. Ketchen (1994) *Phys. Rev. Lett.* **73** 593.
Wollman, D. A., D. J. van Harlingen, W. C. Lee, D. M. Ginsberg and A. J. Leggett (1993) *Phys. Rev. Lett.* **71** 2134.
Zemansky, M. W. and R. H. Dittman (1981) *Heat and Thermodynamics*, McGraw-Hill.

Problems

8.1   We found in Section 8.5 that a charged particle outside a solenoid experiences a transient electric field E and hence a net force $F = qE$ as the current decays in the solenoid. This is initially surprising, as the magnetic field $B = 0$ at all times outside the solenoid, and hence $\nabla \times E = -\partial B/\partial t = 0$ at all times. Why does the charged particle nevertheless experience this transient electric field and force?

8.2   Verify by explicit derivation in Cartesian coordinates that $\nabla \times (\nabla \times j) = \nabla(\nabla \cdot j) - \nabla^2 j$. Using this result and the continuity equation for current density, show that the current density decays inside the plane surface of a superconductor as $|j| = j_0 \exp(-x/\lambda_L)$, where $\mu_0 j_0 \lambda_L = B_0$, λ_L is the London penetration depth, and B_0 is the magnitude of the magnetic field at the superconductor surface. Show also that the magnetic flux penetrating the superconductor per unit length is $B_0 \lambda_L$.

8.3   Combine eqs (8.5) and (8.42) to deduce the temperature dependence of the difference in entropy, $S_N(T) - S_{sc}(T)$, between the normal and superconducting state of a Type I superconductor. Show that this difference is maximised when $T = T_c/\sqrt{3}$.

8.4   The heat capacity C is related to the entropy S by $C = T \partial S/\partial T$. Calculate how the difference in heat capacity between the normal and superconducting states varies with temperature, and hence calculate the magnitude of the discontinuity in the heat capacity at the critical temperature, T_c for aluminium and for niobium. ($T_c(Al) = 1.2\,K$; $T_c(Nb) = 9.2\,K$; $B_c(Al) = 10.5\,mT$; $B_c(Nb) = 206\,mT$).

8.5 A magnetic field B_0 is applied parallel to the surface of a thin super-conducting plate of thickness d which lies in the x–y plane. Taking $z = 0$ at the centre of the plate, show that the magnetic field varies inside the plate as

$$B(z) = B_0 \cosh(z/\lambda_L)/\cosh(d/2\lambda_L)$$

Hence show that if $d \ll \lambda_L$, the magnitude of the mean magnetisation M_{av} will be reduced from B_0/μ_0 to $d^2/(12\lambda_L^2)(B_0/\mu_0)$. It can be shown that this reduction in the average magnetisation leads to an enhancement of the critical field H_c in a thin film, with H_c being proportional to $(\lambda_L/d)H_{c0}$ in a thin film, where H_{c0} is the critical field for a bulk film of the same material.

8.6 Consider the superconducting circuit shown in fig. 8.20, with two identical Josephson junctions in parallel. In the absence of a magnetic field, the phase difference $\Delta\theta$ is the same for the two links, so that the DC Josephson current is given by $I = 2I_0 \sin \Delta\theta$. When a magnetic field is applied, this is no longer the case, so $\Delta\theta_A \neq \Delta\theta_B$. Show by separately integrating around the two sides of the junction that the total difference in phase difference between junction A and junction B equals $q\Phi/\hbar$, where Φ is the total magnetic flux linking the loop, and $q = 2e$ is the charge of a Cooper pair. Show if we set $\Delta\theta = \pi/2$ at junction A, then the total DC Josephson current will vary as $I = I_0(1+\cos(q\Phi/\hbar)) = 2I_0 \cos^2(e\Phi/\hbar)$. This is the superconducting analogue of Young's fringes, with constructive and destructive interference leading to a sinusoidal variation of the DC Josephson current. Show that the period of the oscillations is $\delta\Phi = h/2e = \Phi_0$. How will the DC Josephson current vary if the two junctions are not identical, but instead link a high-T_c and a conventional superconductor, as shown in fig. 8.19?

Appendix A

The variational method with an arbitrary parameter: the H atom

The variational method is one of the key concepts which allows the application of quantum theory to solids. There are few or no circumstances where there is an analytical solution to Schrödinger's equation in a solid. The variational method shows, however, that if we can choose a suitable trial function, such as a linear combination of atomic orbitals, then we can expect to make reasonable estimates of ground state properties and of their variation as a function, for instance, of bond length or ionicity.

We saw at the end of Chapter 1 that the ground state energy estimated by the variational method, $\langle E \rangle$, is always greater than or equal to the true ground state energy, E_0, but that the accuracy with which we can calculate E_0 depends on how well we choose the variational trial function, $f(x)$. The accuracy with which we choose $f(x)$ can be significantly improved by including a free parameter, say γ, in $f(x)$, choosing a function such as $f(x) = e^{-\gamma x}$, and calculating the variational ground state energy $\langle E \rangle$ as a function of γ. When the derivative of $\langle E \rangle$ with respect to γ equals zero, $d\langle E \rangle/d\gamma = 0$, we have (usually) minimised $\langle E \rangle$ and thereby achieved the best possible estimate of the true ground state energy, E_0, for the given trial function, $f(\gamma, x)$, as illustrated in fig. A.1.

This can be very nicely illustrated by considering the electron ground state in the hydrogen atom. The potential $V(r)$ experienced by an electron with charge $-e$ at a distance r from the nucleus is given by $V(r) = -e^2/(4\pi\varepsilon_0 r)$. Because the potential is spherically symmetric, the spherical polar coordinate system is most appropriate for solving the problem. We assume for the trial function a spherically symmetric function, which has its maximum at the origin, $r = 0$, and decays in amplitude for increasing r,

$$f(r) = e^{-\gamma r} \tag{A.1}$$

Using the spherical symmetry, Schrödinger's equation is then given by

$$\left[-\frac{\hbar^2}{2m} \frac{1}{r^2} \frac{d}{dr} \left(r^2 \frac{d}{dr} \right) - \frac{e^2}{4\pi\varepsilon_0 r} \right] \psi(r) = E\psi(r) \tag{A.2}$$

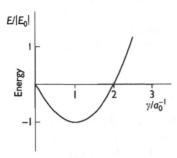

Figure A.1 Variational estimate of the electron ground state energy in a hydrogen atom as a function of the arbitrary parameter γ in the trial wavefunction, $e^{-\gamma r}$. In this case, the lowest variational estimate equals the true ground state energy. The energy scale (vertical axis) is in units of $|E_0|$, the hydrogen ground state binding energy, with the horizontal axis in units of a_0^{-1}, (inverse Bohr radius).

The ground state variational energy can be calculated using eq. (1.37) as

$$\langle E \rangle = \frac{\int_\infty f^*(r)[Hf(r)]\,dV}{\int_\infty f^*(r)f(r)\,dV} \tag{A.3}$$

As the integrands in both the numerator and denominator are spherically symmetric, we can solve eq. (A.3) by integrating outwards over spherical shells of radius r and width dr, for which $dV = 4\pi r^2\,dr$. The denominator in eq. (A.3) is given by

$$\int_{r=0}^{\infty} e^{-2\gamma r}4\pi r^2\,dr = \frac{\pi}{\gamma^3} \tag{A.4}$$

where we use the standard integral $\int_0^\infty e^{-ar}r^n\,dr = n!/a^{n+1}$ while the numerator is given by

$$\int_{r=0}^{\infty} e^{-\gamma r}\left[-\frac{\hbar^2}{2mr^2}\frac{d}{dr}\left(r^2\frac{d}{dr}(e^{-\gamma r})\right) - \frac{e^2}{4\pi\varepsilon_0 r}e^{-\gamma r}\right]4\pi r^2\,dr \tag{A.5}$$

which it can be shown is equal to

$$\frac{\hbar^2\pi}{2m\gamma} - \frac{e^2}{\varepsilon_0(2\gamma)^2} \tag{A.6}$$

with the estimated ground state energy $\langle E \rangle$ then obtained by dividing eq. (A.6) by (A.4) to give

$$\langle E \rangle = \frac{\hbar^2\gamma^2}{2m} - \frac{e^2\gamma}{4\pi\varepsilon_0} \tag{A.7}$$

This expression is reasonable: if we associate γ with the wavenumber k and $1/\gamma = a_0$ with the spatial extent of the trial function then the first term on the right-hand side can be interpreted as the estimated kinetic energy, and the second as the estimated potential energy:

$$\langle E \rangle = \frac{\hbar^2 k^2}{2m} - \frac{e^2}{4\pi\varepsilon_0 a_0} \tag{A.8}$$

where a_0 is referred to as the Bohr radius. To find the minimum estimated ground state energy, we calculate that $d\langle E \rangle / d\gamma = 0$ when $\gamma = e^2 m/(4\pi\varepsilon_0\hbar^2)$, with the Bohr radius $a_0 = 1/\gamma = 0.53\,\text{Å}$. Substituting the calculated γ value in eq. (A.7) gives the estimated electron ground state energy in the hydrogen atom as

$$E_0 = -\frac{me^4}{8\varepsilon_0^2 h^2} = -13.6\,\text{eV} \tag{A.9}$$

In this instance, the calculated minimum variational energy is equal to the ground state energy calculated by solving Schrödinger's equation exactly.

This example demonstrates that the variational method can work very effectively given a suitable choice of starting function, particularly if a free parameter is included in the function. We also observe, as derived in Chapter 1, that the calculated variational energy, $\langle E \rangle \geq E_0$, where E_0 is the true ground state energy, and that in fact $\langle E \rangle = E_0$ only when the variational trial function, $f(\gamma, r)$ equals the true ground state wavefunction, $\psi_0(r)$.

The hydrogen atom and the Periodic Table

The observed structure of the Periodic Table of the elements is due to the ordering of the electron energy levels with increasing atomic number. Although we cannot solve Schrödinger's equation exactly for a multi-electron atom, we can do so for an isolated hydrogen atom, where one electron orbits a positively charged nucleus. The calculated ordering of the hydrogen atom ground and excited state energy levels can then account for the trends observed in the Periodic Table. We first outline here the solution of the hydrogen atom Schrödinger equation and then apply it to explain some of the main trends in the Periodic Table. A more detailed solution of the hydrogen atom can be found in almost all quantum mechanics textbooks (e.g. Davies and Betts 1994; McMurry 1993; Schiff 1968).

The potential $V(r)$ experienced by an electron with charge $-e$ at a distance r from the hydrogen nucleus is given by

$$V(r) = \frac{-e^2}{4\pi\varepsilon_0 r} \tag{B.1}$$

Because the potential is spherically symmetric, the spherical polar coordinate system (r, θ, ϕ) is most appropriate for solving the problem. Schrödinger's equation is given by

$$\left[-\frac{\hbar^2}{2m}\nabla^2 - \frac{e^2}{4\pi\varepsilon_0 r} \right] \psi(r) = E\psi(r) \tag{B.2}$$

with the operator ∇^2 given in spherical polars by

$$\nabla^2 = \frac{1}{r^2}\frac{\partial}{\partial r}\left(r^2\frac{\partial}{\partial r}\right) + \frac{1}{r^2 \sin\theta}\frac{\partial}{\partial\theta}\left(\sin\theta\frac{\partial}{\partial\theta}\right) + \frac{1}{r^2 \sin^2\theta}\frac{\partial^2}{\partial\phi^2} \tag{B.3}$$

We separate the radial and angular parts of Schrödinger's equation by substituting $\psi(r, \theta, \phi) = R(r)\Theta(\theta)\Phi(\phi)$ into eq. (B.2), dividing through by

ψ and re-arranging to give

$$\frac{1}{R}\frac{d}{dr}\left(r^2\frac{dR}{dr}\right) + \frac{2mr^2}{\hbar^2}[E - V(r)]$$

$$= -\frac{1}{\Theta\sin\theta}\frac{\partial}{\partial\theta}\left(\sin\theta\frac{\partial\Theta}{\partial\theta}\right) - \frac{1}{\Phi\sin^2\theta}\frac{\partial^2\Phi}{\partial\phi^2} \tag{B.4}$$

Because the left-hand side of eq. (B.4) depends only on r, and the right-hand side contains two terms, one of which depends only on θ and the other on θ and ϕ, both sides must be equal to a constant, which we call λ. It is important to note that the separation of variables in eq. (B.4) can be carried out for any spherically symmetric potential; not just the hydrogen atom potential of eq. (B.1).

The right-hand side of eq. (B.4), describing the angular variation of the wavefunction, can be further separated into two simpler equations by introducing an additional constant, m^2, such that

$$\frac{d^2\Phi}{d\phi^2} + m^2\Phi = 0 \tag{B.5}$$

$$\frac{1}{\sin\theta}\frac{d}{d\theta}\left(\sin\theta\frac{d\Theta}{d\theta}\right) + \left(\lambda - \frac{m^2}{\sin^2\theta}\right)\Theta = 0 \tag{B.6}$$

The first of these two equations (B.5) can be solved at once; we find

$$\Phi(\phi) = \frac{1}{\sqrt{2\pi}}e^{im\phi} \tag{B.7}$$

where m must be an integer in order that Φ is single-valued, with $\Phi(2\pi) = \Phi(0)$.

The second equation is generally solved by making the change of variables $z = \cos\theta$, so that eq. (B.6) becomes

$$\frac{d}{dz}\left[(1 - z^2)\frac{dP}{dz}\right] + \left(\lambda - \frac{m^2}{1 - z^2}\right)P = 0 \tag{B.8a}$$

which we can rewrite as

$$(1 - z^2)\frac{d}{dz}\left[(1 - z^2)\frac{dP}{dz}\right] + \left[\lambda(1 - z^2) - m^2\right]P = 0 \tag{B.8b}$$

where $P(z) = \Theta(\cos\theta)$. Equation (B.8) is well known to mathematicians as the associated Legendre equation. Its solution may readily be found using the series method: full details are given, for example, in Davies and Betts (1994). The allowed solutions are polynomial functions containing a finite number of terms. Here we need only note that allowed solutions will only exist if the coefficient of each power of z in the polynomial is identically zero when substituted into eq. (B.8).

If the leading term of the polynomial P is z^n, substitution into eq. (B.8b) shows that the leading power in the differential equation is z^{n+2}, with coefficient $n(n-1) - \lambda^2$. This must vanish, so $\lambda = n(n-1)$. It is conventional to put $n = l + 1$, and so to write

$$\lambda = l(l+1) \tag{B.9}$$

Further detailed analysis shows that allowed solutions can only exist when $l(l+1) > m^2$, which requires

$$-l \le m \le l \tag{B.10}$$

so that there are $2l + 1$ allowed values of m for each value of l.

We now substitute eq. (B.9) for λ back into eq. (B.4) to derive the radial Schrödinger equation

$$-\frac{1}{r^2}\frac{\hbar^2}{2m}\frac{d}{dr}\left(r^2\frac{dR}{dr}\right) + \left(\frac{\hbar^2}{2m}\frac{l(l+1)}{r^2} + V(r)\right)R(r) = ER(r) \tag{B.11}$$

The l-dependent term may be written as $Q^2/2mr^2$, and is the quantum counterpart of the classical 'centrifugal' potential barrier $Q^2/2mr^2$ encountered for example in the Kepler problem of planetary orbits, where $Q = mr^2\omega$ is the angular momentum of the orbiting particle. We have thus shown that the angular momentum is quantised in a spherically symmetric potential, with the magnitude squared of the angular momentum $Q^2 = \hbar^2 l(l+1)$, where l is an integer. Further analysis reveals that if we quantise along a particular direction (e.g. along the z-axis) then the angular momentum component along that axis is also quantised, with the component Q_z projected onto that axis equal to

$$Q_z = \hbar m, \qquad |m| \le l \tag{B.12}$$

We note that eq. (B.11) depends on the total angular momentum through the term containing $l(l+1)$, but does *not* depend on m, the angular momentum component along the quantisation axis. This is to be expected. The energy should not depend on the orientation of the z-axis in a spherically symmetric potential.

Equation (B.11) can be solved for the hydrogen atom using standard mathematical techniques, again described in many quantum mechanics texts. To avoid having to work with a large number of constants ($e, m, \hbar$, etc.) we introduce a change of variables

$$\rho = \alpha r; \quad \alpha^2 = \frac{-8mE}{\hbar^2}; \quad \beta = \frac{2me^2}{4\pi\varepsilon_0\alpha\hbar^2} \tag{B.13}$$

whereupon eq. (B.11) simplifies to

$$\left[\frac{1}{\rho^2}\frac{d}{d\rho}\left(\rho^2\frac{d}{d\rho}\right) + \frac{\beta}{\rho} - \frac{1}{4} - \frac{l(l+1)}{\rho^2}\right]R(\rho) = 0 \tag{B.14}$$

It can be shown that solutions to this equation can be written in the form $F(\rho)\exp(-\rho/2)$, where the polynomial F satisfies the differential equation

$$\rho^2\frac{d^2F}{d\rho^2} + (2\rho - \rho^2)\frac{dF}{d\rho} + [\rho(\beta - 1) - l(l + 1)]F = 0 \tag{B.15}$$

In this case, if the leading term of the polynomial F is ρ^k, substitution into eq. (B.15) shows that the leading power of the differential equation is ρ^{k+1}, with coefficient $\beta - k - 1$. This term cannot cancel against lower order terms in the polynomial, as all these have powers ρ^k or less; so to satisfy eq. (B.15) we must have $\beta = k + 1$, $k = 0, 1, 2, \ldots$, and also $k \geq l$. This is more usually written as

$$\beta = n, \qquad n = 1, 2, 3, \ldots \tag{B.16}$$

with $n > l$. Combining eqs (B.13) and (B.16) we see that the hydrogen atom energy levels are then given by

$$E = E_n = -\frac{me^4}{2(4\pi\varepsilon_0)^2\hbar^2n^2} \tag{B.17}$$

To summarise, we have thus deduced three quantum numbers associated with each allowed energy state in the hydrogen atom, namely n, l and m. The energy of each allowed hydrogen state depends only on n, which is therefore referred to as the *principal* quantum number, with l referred to as the angular momentum or *orbital* quantum number and m as the *magnetic* quantum number. In addition, each electron has an intrinsic *spin*, which can take the values of $s_z = \pm\frac{1}{2}$. The quantum number names, symbols, and allowed values are summarised in Table B.1, along with the physical property to which they are related.

The energy levels in eq. (B.17) do not depend on l, m or s_z. Therefore, each energy level in the hydrogen atom has a multiple degeneracy, with the number N of degenerate states depending on the number of allowed values of l and m for each principal quantum number n:

$$N = 2\sum_{l=0}^{n-1}(2l + 1) = 2n^2 \tag{B.18}$$

where the factor 2 comes from the existence of two spin states. Because all states with the same principal quantum number n have the same energy, we therefore say those states belong to the nth *shell* of states.

Table B.1 Details of quantum numbers associated with energy levels of an isolated atom

Name	Symbol	Values allowed	Physical property
Principal	n	$n = 1, 2, 3, \ldots$	Determines radial extent and energy
Orbital	l	$l = 0, 1, \ldots, (n-1)$	Angular momentum and orbit shape
Magnetic	m	$-l, -l+1, \ldots, l-1, l$	Projection of orbital angular momentum along quantisation axis
Spin	s_z	$+\frac{1}{2}$ and $-\frac{1}{2}$	Projection of electron spin along quantisation axis

Table B.2 Spectroscopic labels associated with different orbital quantum numbers (atomic subshells) in an isolated atom

Orbital quantum number l	0	1	2	3	4	5
Spectroscopic label	s	p	d	f	g	h

The degeneracy we have found for states with the same principal quantum number n, but different orbital quantum numbers l, is an 'accidental' consequence of the hydrogen atom potential $V(r)$, which varies as $1/r$. This 'accidental' l-degeneracy is removed when most other central potentials are used in (B.11), including the potential of any multi-electron atom. Each shell of states with particular principal quantum number n therefore breaks up into a set of subshells in the multi-electron atom, with a different orbital quantum number l associated with each subshell. Historically, the states in different subshells were identified using spectroscopic techniques, and the different subshells were given the spectroscopic labels shown in Table B.2. States with $n = 1$ and $l = 0$ are referred to as 1s states, while states with $n = 2$ and $l = 0$ and 1 are referred to as 2s and 2p states, respectively.

We recall from eq. (B.11) that the effect of increasing angular momentum (increasing l) is described by an increasingly strong centrifugal potential barrier (proportional to $l(l+1)/r^2$) which pushes the electron away from the nucleus. As a consequence, the 2s wavefunction (with $l = 0$) will have larger amplitude close to the nucleus than does the 2p wavefunction (with $l = 1$). The 2s states therefore experience on average a stronger attractive potential, and so will be at a lower energy than the 2p states. The energies of the different electron states in a multi-electron atom are clearly affected by the presence of other electrons. The 1s orbital will always have lowest energy and, because it largely experiences the full nuclear attraction (proportional to Ze for atomic number Z), its binding energy will be close to Z^4 times the binding energy of the 1s hydrogen state. The 1s states will then partially screen the higher lying levels, modifying their energies accordingly.

1 H																	2 He
3 Li	4 Be											5 B	6 C	7 N	8 O	9 F	10 Ne
11 Na	12 Mg											13 Al	14 Si	15 P	16 S	17 Cl	18 Ar
19 K	20 Ca	21 Sc	22 Ti	23 V	24 Cr	25 Mn	26 Fe	27 Co	28 Ni	29 Cu	30 Zn	31 Ga	32 Ge	33 As	34 Se	35 Br	36 Kr
37 Rb	38 Sr	39 Y	40 Zr	41 Nb	42 Mo	43 Tc	44 Ru	45 Rh	46 Pd	47 Ag	48 Cd	49 In	50 Sn	51 Sb	52 Te	53 I	54 Xe
55 Cs	56 Ba	57 La	72 Hf	73 Ta	74 W	75 Re	76 Os	77 Ir	78 Pt	79 Au	80 Hg	81 Tl	82 Pb	83 Bi	84 Po	85 At	86 Rn
87 Fr	88 Ra	89 Ac															

58 Ce	59 Pr	60 Nd	61 Pm	62 Sm	63 Eu	64 Gd	65 Tb	66 Dy	67 Ho	68 Er	69 Tm	70 Yb	71 Lu
90 Th	91 Pa	92 U	93 Np	94 Pu	95 Am	96 Cm	97 Bk	98 Cf	99 Es	100 Fm	101 Md	102 No	103 Lr

Figure B.1 Periodic Table of the elements. The atomic number of each element is given in the top left hand corner of its box, with the chemical symbol in the box centre. The elements are arranged in columns predominantly reflecting the order in which electronic subshells are filled: s states ($l = 0$) filling in the first two columns; p-states ($l = 1$) in the six right-hand columns, with d-state ($l = 2$) filling in columns 3–12, and f-state filling ($l = 3$) indicated in the 'footnote' to the Table.

Figure B.1 presents the Periodic Table of the elements. The analysis we have presented here accounts for most of the trends observed in the Periodic Table, which we assume and use throughout this book:

1 In a many-electron atom, the lowest energy states typically experience a very strong attractive potential due to the positively charged nucleus. These states therefore have a much larger binding energy than the highest filled levels. They are referred to as *core* states. They are highly localised and make no direct contribution to bonding in a solid.

2 The core electrons screen the attractive potential seen by the higher filled states. We can compare, for example, carbon (C, $Z = 6$) with silicon (Si, $Z = 14$) and germanium (Ge, $Z = 32$). For C, the two filled 1s states at least partly cancel the attraction due to two of the six nuclear protons, so that the 2s and 2p states experience an average attractive potential equivalent to of order four protons. In Si, the filled 2s and 2p states also contribute to the screening, so that the 3s and 3p states again experience an average attractive potential equivalent to of order four protons. Likewise in Ge, the filled $n = 3$ shell (3s, 3p, 3d states) contributes to screening, leaving a similar net attractive potential to

that found in C and Si. Hence, the outermost *valence* states have similar character for the three elements.

3   We have already noted that because of the angular-momentum-related repulsive barrier in eq. (B.11) the valence s states will always lie below the valence p states.

4   When we again compare C, Si and Ge, we note that in each case the outermost valence states must be orthogonal to the core states. Because there are more core electrons as we go further down the Periodic Table, the outermost electrons tend to have a larger spatial extent in Ge than in Si than in C, so that the atom size increases going down a row of the Periodic Table. Likewise, the valence electrons have a larger binding energy in C than in Si than in Ge.

5   If we now look at a set of elements in the same row of the Periodic Table, such as aluminium (Al, $Z = 13$), silicon (Si, $Z = 14$) and phosphorus (P, $Z = 15$), we note that in each case there are 10 'core' electrons screening the nuclear attraction, thereby leaving a net attraction of order three, four and five protons for Al, Si, and P, respectively. There is also, of course, a repulsive interaction between the increasing number of valence electrons in each atom, but the increasing net nuclear attraction dominates, so that the valence state binding energy tends to increase with increasing atomic number, giving what is referred to as increasing electronegativity across a row of the Periodic Table.

6   The ordering of some of the higher lying subshells does not always follow the main shell order. Thus all subshells up to the 3p subshell are filled for argon (Ar, $Z = 18$). However, the 4s subshell lies below the 3d subshell for potassium (K, $Z = 19$), so that the 4s subshell first starts to fill with increasing Z, followed by the 3d subshell and then the 4p subshell.

7   Finally, we do not discuss here the order in which different states are filled within a partly filled subshell of an atom. This is discussed in Chapter 6, where we introduce Hund's rules. They were originally derived empirically (later justified by careful quantum mecahnical analysis) and describe the order in which states with different values of m and s_z are occupied. This is generally not of relevance when considering bonding in solids, as in Chapters 2–5, but becomes of key significance when considering the magnetic properties of atoms and solids in Chapters 6 and 7.

References

Davies, P. C. W. and D. S. Betts (1994) *Quantum Mechanics*, 2nd edn, Nelson Thornes, Cheltenham.

McMurry, S. M. *Quantum Mechanics*, Addison-Wesley.

Schiff, L. I. (1968) *Quantum Mechanics*, 3rd edn, McGraw-Hill, Tokyo.

Appendix C

First and second order perturbation theory

C.1 Introduction to perturbation theory

There are remarkably few potential energy functions for which it is possible to find an exact, analytical solution of Schrödinger's equation. The main cases for which exact solutions exist include the infinite and the finite square well, the hydrogen atom (discussed in Appendix A) and the simple harmonic oscillator, for which the potential energy varies as $V(r) = \frac{1}{2}kr^2$. Because of their analytical solutions, all of these potentials get used extensively throughout this and all quantum mechanics books. We have, however, seen that even in cases where a potential has no exact solution, we can make a very good estimate of the ground and first excited state energies by using the variational method, where we guess the form $f(x)$ of the ground state wavefunction and then calculate the estimated energy using eq. (1.37).

This is by no means the only approximation method which is useful in quantum mechanics. There are many problems where the full Hamiltonian H cannot be solved exactly, but where H can be written as the sum of two parts,

$$H = H_0 + H' \tag{C.1}$$

where the first part H_0 is of sufficiently simple structure that its Schrödinger equation can be solved exactly, while the second part H' is small enough that it can be regarded as a *perturbation* on H_0. An example of such a problem is a hydrogen atom in an applied electric field (fig. C.1) for which

$$H\psi(r) = \left(-\frac{\hbar^2}{2m}\nabla^2 - \frac{e^2}{4\pi\varepsilon_0 r} + eEz \right)\psi(r) \tag{C.2}$$

where H_0 is the hydrogen atom Hamiltonian

$$H_0\psi(r) = \left(-\frac{\hbar^2}{2m}\nabla^2 - \frac{e^2}{4\pi\varepsilon_0 r} \right)\psi(r) \tag{C.3}$$

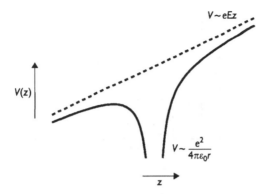

$V \sim eEz$

$V(z)$

$V \sim -\dfrac{e^2}{4\pi\varepsilon_0 r}$

z

Figure C.1 Solid line: the variation in the potential seen across a hydrogen atom due to an electric field E applied along the z-direction. (Dashed line shows the contribution to the total potential due to the electric field.)

while H' is the change in potential due to the applied field

$$H' = eEz \tag{C.4}$$

In such cases, we can often make a very good estimate of both the ground and excited state energies by first solving the Hamiltonian H_0 exactly and then using an approximation method known as perturbation theory to estimate how H' shifts the energy levels from their H_0 values. In the following sections, we first describe the principles of first order perturbation theory, which is closely related to the variational method, and then consider second order perturbation theory. Second order perturbation theory forms the basis for the $k \cdot p$ description of crystal band structure, which we derive in Appendix E and apply in Chapters 4 and 5.

C.2   First order perturbation theory

First order perturbation theory is closely related to the variational method. To estimate the energy states of the Hamiltonian, $H = H_0 + H'$, we need to make the best possible guess for the wavefunctions of H. We know and can solve

$$H_0 \psi_k^{(0)}(r) = E_k^{(0)} \psi_k^{(0)}(r) \tag{C.5}$$

where $E_k^{(0)}$ is the energy and $\psi_k^{(0)}(r)$ the normalised wavefunction of the kth energy state of H_0. We choose $\psi_k^{(0)}(r)$ as the trial wavefunction for the kth state of the full Hamiltonian, H, and then use the variational method

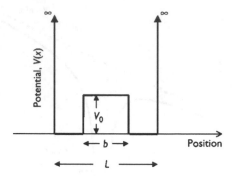

Figure C.2 Infinite square well of width L, with a barrier of height V_0 and width b added in its centre.

to estimate the energy of that state, $W_k^{(1)}$, as

$$W_k^{(1)} = \frac{\int dV \psi_k^{*(0)}(r) H \psi_k^{(0)}(r)}{\int dV \psi_k^{*(0)}(r) \psi_k^{(0)}(r)} \tag{C.6}$$

where the denominator in eq. (C.6) is equal to 1. Replacing H by $H_0 + H'$ and splitting the integral in eq. (C.6) into two parts gives

$$W_k^{(1)} = \int dV \psi_k^{*(0)}(r) H_0 \psi_k^{(0)}(r) + \int dV \psi_k^{*(0)}(r) H' \psi_k^{(0)}(r)$$

$$= E_k^{(0)} + \int dV \psi_k^{*(0)}(r) H' \psi_k^{(0)}(r) \tag{C.7}$$

The energy levels $E_k^{(0)}$ of the Hamiltonian H_0 thus provide the zeroth-order guess for the energy levels of the full Hamiltonian H, while $E_k^{(1)} = \int dV \psi_k^{*(0)}(r) H' \psi_k^{(0)}(r)$ gives the first order correction to the estimated energy.

C.2.1 Example: double square well with infinite outer barriers

To illustrate the application of first order perturbation theory, we consider an infinite square well in the region $0 < x < L$, to which is added a potential barrier of height V_0 and width b between $L/2 - b/2$ and $L/2 + b/2$ (fig. C.2). The energy levels $E_n^{(0)}$ and wavefunctions $\psi_n^{(0)}(x)$ of the unperturbed well are given by (eq. (1.11))

$$E_n^{(0)} = \frac{h^2 n^2}{8mL^2} \tag{C.8a}$$

and

$$\psi_n^{(0)}(x) = \sqrt{\frac{2}{L}} \sin\left(\frac{n\pi x}{L}\right) \tag{C.8b}$$

Substituting eq. (C.8b) in eq. (C.7), the estimated first order shift in the energy levels is given by

$$
\begin{aligned}
E_n^{(1)} &= \frac{2}{L} \int_{L/2-b/2}^{L/2+b/2} V_0 \sin^2\left(\frac{n\pi x}{L}\right) dx \\
&= V_0 \left[\frac{b}{L} + \frac{(-1)^{n+1}}{n\pi} \sin\left(\frac{n\pi b}{L}\right)\right]
\end{aligned} \tag{C.9}
$$

That is, $E_n^{(1)}$ varies linearly with the height of the perturbing potential, V_0. The solid lines in fig. C.3 show how the true energy levels vary with the barrier height V_0 in the case where the barrier width b is half of the total infinite well width, $b = L/2$, while the straight dashed lines show the estimated variation using first order perturbation theory. It can be seen that first order perturbation theory is indeed useful for small perturbations, but rapidly becomes less accurate as V_0 increases. The accuracy of the perturbation estimate can, however, be extended to larger values of V_0 by going to second order perturbation theory, which will then introduce further correction terms to the estimated energy of order V_0^2.

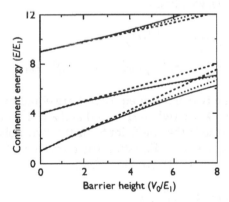

Figure C.3 Variation of the confined state energy for the three lowest energy levels in an infinite square well, as a function of the magnitude of the perturbing potential, V_0 introduced in fig. C.2. The barrier width b is set equal to half of the total infinite well width, $b = L/2$, and V_0 is plotted in units of $h^2/8mL^2$ (the ground state energy, E_1). Solid line: exact solution; dashed (straight) lines: using first order perturbation theory; dotted (parabolic) lines: using second order perturbation theory.

C.3   Second order perturbation theory

In second order perturbation theory, we first estimate how the wavefunction of the kth state is changed by the additional potential H', and then effectively use the modified wavefunction as input to a variational method, to get an improved, second order, estimate of the energy.

The assumption that the perturbation H' is small allows us to expand the perturbed wavefunction and energy value as a power series in H'. This is most conveniently accomplished by introducing a parameter λ such that the zeroth, first, etc. powers of λ correspond to the zeroth, first, etc. orders of the perturbation calculation. If we replace H' by $\lambda H'$ in eq. (C.1) then H is equal to the full Hamiltonian, $H = H_0 + H'$, when $\lambda = 1$, while H equals the unperturbed Hamiltonian, $H = H_0$ when $\lambda = 0$.

We can express both the wavefunctions $\psi_k(r)$ and energy levels E_k of the Hamiltonian H as a power series in λ. For second order perturbation theory, we require the first order approximation to $\psi_k(r)$ and the second order approximation to the energy W_k, and so write

$$\psi_k(r) = \psi_k^{(0)}(r) + \lambda \psi_k^{(1)}(r) \tag{C.10a}$$

and

$$W_k^{(2)} = E_k^{(0)} + \lambda E_k^{(1)} + \lambda^2 E_k^{(2)} \tag{C.10b}$$

Substituting the wavefunction and energy described by eq. (C.10) into the wave equation (C.1), we obtain

$$(H_0 + \lambda H') \left(\psi_k^{(0)}(r) + \lambda \psi_k^{(1)}(r) \right)$$
$$= \left(E_k^{(0)} + \lambda E_k^{(1)} + \lambda^2 E_k^{(2)} \right) \left(\psi_k^{(0)}(r) + \lambda \psi_k^{(1)}(r) \right) \tag{C.11}$$

We assume that the two sides of eq. (C.11) are equal to each other for all values of λ between $\lambda = 0$ (when $H = H_0$) and $\lambda = 1$ (when $H = H_0 + H'$). This can only be true if the coefficients of the two polynomials in λ are identical for each power of λ. Equating the terms for different powers of λ on the two sides of eq. (C.11), we then obtain

$$\lambda^0 : \quad H_0 \psi_k^{(0)}(r) = E_k^{(0)} \psi_k^{(0)}(r) \tag{C.12a}$$

$$\lambda^1 : \quad H' \psi_k^{(0)}(r) + H_0 \psi_k^{(1)}(r) = E_k^{(0)} \psi_k^{(1)}(r) + E_k^{(1)} \psi_k^{(0)}(r) \tag{C.12b}$$

$$\lambda^2 : \quad H' \psi_k^{(1)} = E_k^{(2)} \psi_k^{(0)}(r) + E_k^{(1)} \psi_k^{(1)}(r) \tag{C.12c}$$

The first of these equations (C.12a), is just the Schrödinger equation for the unperturbed Hamiltonian, H_0, while the second equation (C.12b) can be used to calculate the first order correction to the energy levels, $E_k^{(1)}$, and wavefunctions, $\psi_k^{(1)}$. Finally, we can substitute the results of eq. (C.12b) into the third equation (C.12c), to determine the second order change, $E_k^{(2)}$ in the energy levels.

We wish first to consider the form of $\psi_k^{(1)}(r)$, the *change* to $\psi_k^{(0)}(r)$ due to the perturbation H'. We described in Chapter 1 how any function $f(r)$ can be written as a linear combination of the complete set of states, $\psi_n^{(0)}(r)$. The *change* in the wavefunction, $\psi_k^{(0)}(r)$, due to the perturbation H' will involve mixing ('adding') other states into $\psi_k^{(0)}(r)$, so we expect that $\psi_k^{(1)}(r)$ can be written as a linear combination of all the other wavefunctions

$$\psi_k^{(1)}(r) = \sum_{n \neq k} a_{kn} \psi_n^{(0)}(r) \tag{C.13}$$

where a_{kn} is the amplitude which the nth state contributes to the modification of the kth wavefunction. Substituting eqs (C.12a) and (C.13) into eq. (C.12b), and rearranging, we obtain

$$\sum_{n \neq k} \left(H_0 - E_k^{(0)} \right) a_{kn} \psi_n^{(0)}(r) = \left(E_k^{(1)} - H' \right) \psi_k^{(0)}(r) \tag{C.14}$$

which, as $H_0 \psi_n^{(0)}(r) = E_n^{(0)} \psi_n^{(0)}(r)$, reduces to

$$\sum_{n \neq k} a_{kn} \left(E_n^{(0)} - E_k^{(0)} \right) \psi_n^{(0)}(r) = \left(E_k^{(1)} - H' \right) \psi_k^{(0)}(r) \tag{C.15}$$

We can use eq. (C.15) to evaluate the two first order corrections, $E_k^{(1)}$ and $\psi_k^{(1)}(r)$. We first multiply both sides of eq. (C.15) by $\psi_k^{*(0)}(r)$, and integrate over all space to find $E_k^{(1)}$:

$$\sum_{n \neq k} a_{kn} \left(E_n^{(0)} - E_k^{(0)} \right) \int dV \, \psi_k^{*(0)}(r) \psi_n^{(0)}(r)$$

$$= \int dV \, \psi_k^{*(0)}(r) \left(E_k^{(1)} - H' \right) \psi_k^{(0)}(r) \tag{C.16}$$

The left hand side of eq. (C.16) is identically zero, because the wavefunctions $\psi_k^{(0)}(r)$ and $\psi_n^{(0)}(r)$ are orthogonal, and we can rearrange the right-hand side to give the same result for the first order energy correction as in the previous section:

$$E_k^{(1)} = \int dV \, \psi_k^{*(0)}(r) H' \psi_k^{(0)}(r) \tag{C.7}$$

We use the same technique to calculate the coefficients a_{km} from eq. (C.13), multiplying both sides of eq. (C.15) by $\psi_m^{*(0)}(r)$, and integrating over all space:

$$\sum_{n \neq k} a_{kn} \left(E_n^{(0)} - E_k^{(0)} \right) \int dV \psi_m^{*(0)}(r) \psi_n^{(0)}(r)$$

$$= \int dV \, \psi_m^{*(0)}(r) \left(E_k^{(1)} - H' \right) \psi_k^{(0)}(r) \tag{C.17}$$

Most of the terms in this equation are again equal to zero, so that it reduces to

$$a_{km} \left(E_m^{(0)} - E_k^{(0)} \right) = - \int dV \, \psi_m^{*(0)}(r) H' \psi_k^{(0)}(r)$$

$$= -\langle \psi_m^{(0)} | H' | \psi_k^{(0)} \rangle \tag{C.18}$$

where $\langle \psi_m^{(0)} | H' | \psi_k^{(0)} \rangle$ is a commonly used 'shorthand' notation for the integral $\int dV \, \psi_m^{*(0)}(r) H' \psi_k^{(0)}(r)$ (see Appendix D). Substituting eq. (C.18) back into eq. (C.13), the calculated first order change in the kth wavefunction, $\psi_k^{(1)}(r)$, is then given by

$$\psi_k^{(1)}(r) = \sum_{n \neq k} \frac{\langle \psi_n^{(0)} | H' | \psi_k^{(0)} \rangle}{E_k^{(0)} - E_n^{(0)}} \psi_n^{(0)}(r) \tag{C.19}$$

We now substitute eq. (C.13) directly into eq. (C.12c) to calculate $E_k^{(2)}$, the second order correction to the kth energy level. Taking the terms involving $\psi_k^{(1)}(r)$ to the left-hand side of eq. (C.12c), and substituting eq. (C.13) for $\psi_k^{(1)}(r)$ we find

$$\sum_{n \neq k} \left(H' - E_k^{(1)} \right) a_{kn} \psi_n^{(0)}(r) = E_k^{(2)} \psi_k^{(0)}(r) \tag{C.20}$$

This time, we multiply both sides of eq. (C.20) by $\psi_k^{*(0)}(r)$, and integrate over all space to find

$$\sum_{n \neq k} \int dV \, \psi_k^{*(0)}(r) \left(H' - E_k^{(1)} \right) a_{kn} \psi_n^{(0)}(r) = E_k^{(2)} \int dV \, \psi_k^{*(0)}(r) \psi_k^{(0)}(r)$$

$$\tag{C.21}$$

We again use the orthogonality property of the wavefunctions, to find that the second order energy correction, $E_k^{(2)}$, is given (in the 'shorthand' notation of Appendix D) by

$$E_k^{(2)} = \sum_{n \neq k} a_{kn} \langle \psi_k^{(0)} | H' | \psi_n^{(0)} \rangle$$

$$= \sum_{n \neq k} \frac{\langle \psi_n^{(0)} | H' | \psi_k^{(0)} \rangle}{E_k^{(0)} - E_n^{(0)}} \langle \psi_k^{(0)} | H' | \psi_n^{(0)} \rangle \qquad (C.22)$$

with the perturbed energy levels, $W_k^{(2)}$, then given to second order by

$$W_k^{(2)} = E_k^{(0)} + \langle \psi_k^{(0)} | H' | \psi_k^{(0)} \rangle + \sum_{n \neq k} \frac{|\langle \psi_n^{(0)} | H' | \psi_k^{(0)} \rangle|^2}{E_k^{(0)} - E_n^{(0)}} \qquad (C.23)$$

There are two consequences of eq. (C.23) which are worth noting:

1 The first order correction, $\langle \psi_k^{(0)} | H' | \psi_k^{(0)} \rangle$, describes the change in energy due to the change in potential, H', seen by the unperturbed wavefunction, $\psi_k^{(0)}(r)$, and can shift the energy W_k upwards or downwards, depending on the potential change. Thus, in the example we considered earlier, because the potential V_0 increased in the barrier region between $L/2 - b/2$ and $L/2 + b/2$, the first order correction shifted *all* the levels upwards, with the size of the shift depending on the probability of finding the kth state in the barrier region.

2 The second order correction, $E_k^{(2)}$, describes the change in energy due to mixing between states in the perturbing potential. The numerator of eq. (C.22), $|\langle \psi_n^{(0)} | H' | \psi_k^{(0)} \rangle|^2$, is always positive, while the denominator is negative if the nth state is above the kth state ($E_n^{(0)} > E_k^{(0)}$) and positive if $E_n^{(0)}$ is below $E_k^{(0)}$. Hence, mixing between any two states always tends to push them apart, increasing their energy separation. Also, the effect of mixing with higher levels always leads to a downward second order shift in the ground state energy.

In deriving eq. (C.23), we made a couple of assumptions which are not always true: first, we assumed that we were dealing only with *discrete* energy levels (bound states), and second, we assumed that no other state was degenerate, that is, had the same energy as $E_k^{(0)}$. If, however, we are dealing with degenerate states, so that say $E_m^{(0)} = E_k^{(0)}$, then the above analysis needs to be modified, to avoid the possibility of the denominator of the second order term, $E_k^{(0)} - E_m^{(0)}$, becoming equal to zero. Details of the modified analysis can be found in more advanced textbooks, such as Schiff's *Quantum Mechanics* (pp. 248ff.). We do not derive the modified analysis here, but implicitly use it as necessary in the main text.

C.3.1   Example: double square well with infinite outer barriers

To illustrate the application of second order perturbation theory we return to the problem we considered earlier of an infinite square well of width L to whose centre is added a barrier of height V_0 and width b (fig. C.2). The matrix element $\langle \psi_m^{(0)} | H' | \psi_n^{(0)} \rangle$ describing the mixing between the mth and nth state is given by

$$\langle \psi_m^{(0)} | H' | \psi_n^{(0)} \rangle = \frac{2}{L} \int_{L/2-b/2}^{L/2+b/2} V_0 \sin\left(\frac{m\pi x}{L}\right) \sin\left(\frac{n\pi x}{L}\right) dx \qquad \text{(C.24)}$$

This integral can be solved using the identity $\sin \alpha \sin \beta = \frac{1}{2}(\cos(\alpha - \beta) - \cos(\alpha + \beta))$, to give

$$\langle \psi_m^{(0)} | H' | \psi_n^{(0)} \rangle$$

$$= \frac{2V_0}{\pi} \left[\frac{\sin\left((n - m)\pi b/2L\right)}{n - m} - (-1)^n \frac{\sin\left((n + m)\pi b/2L\right)}{n + m} \right] \quad n + m \text{ even}$$

$$= 0 \qquad\qquad\qquad\qquad\qquad\qquad\qquad\qquad\qquad\qquad n + m \text{ odd}$$

$$\text{(C.25)}$$

The dotted (parabolic) lines in fig. C.3 show how the calculated energy levels vary with barrier height V_0 in second order perturbation theory. It can be seen that going to second order gives a useful improvement in the range of V_0 values over which perturbation theory applies. You might then consider extending perturbation theory to third or even higher orders: however, a law of diminishing returns rapidly sets in, and in practice no advantage is gained by extending perturbation theory beyond the second order.

Reference

Schiff, L. I. (1968) *Quantum Mechanics*, 3rd edn, McGraw-Hill, Tokyo.

Appendix D

Dirac notation

In many quantum mechanics derivations, we need to evaluate integrals involving the product of the complex conjugate of a wavefunction, $\phi_m^*(r)$, times an operator, say H, operating on another wavefunction, say $\phi_n(r)$. The integral I is then given by

$$I = \int_{\text{All space}} dV\, \phi_m^*(r) H\, \phi_n(r) \tag{D.1}$$

If we have such integrals on both sides of an equation, then the equation becomes very long when written out in full, as we see, for example, for eq. (3.22a) in Chapter 3. It is, therefore, useful to introduce a short-hand notation, which conveys the same information as eq. (3.22a), but in a more compact form. We do so using Dirac notation, where we define the wavefunction and its complex conjugate by

$$|\phi_m\rangle \equiv \phi_m(r) \tag{D.2a}$$

and

$$\langle\phi_m| \equiv \phi_m^*(r) \tag{D.2b}$$

$\langle\phi_m|$ and $|\phi_m\rangle$ are referred to as a *bra* and *ket*, respectively. When written separately, as in eq. (D.2), they provide a not-very-shorthand way of denoting the wavefunction and its complex conjugate. Likewise, we have the 'shorthand' notation whereby $H|\phi_m\rangle$ denotes the operator H acting on the wavefunction $\phi_m(r)$; that is,

$$H|\phi_m\rangle \equiv H\phi_m(r) \tag{D.3}$$

The Dirac notation becomes most useful when we need to write down overlap integrals, such as that in eq. (3.22a). We define $\langle\phi_m|H|\phi_n\rangle$ as the integral of the product of $\phi_m^*(r)$ times H times $\phi_n(r)$, while $\langle\phi_m|\phi_n\rangle$ is defined as the integral of the product of $\phi_m^*(r)$ times $\phi_n(r)$; that is,

$$\langle\phi_m|H|\phi_n\rangle \equiv \int_{\text{All space}} dV\, \phi_m^*(r)H\,\phi_n(r) \tag{D.4}$$

and

$$\langle\phi_m|\phi_n\rangle \equiv \int_{\text{All space}} dV\, \phi_m^*(r)\phi_n(r) \tag{D.5}$$

We introduce Dirac notation for the first time in Chapter 3, using it there and later mainly in cases where it significantly shortens the length of equations.

Appendix E

Bloch's theorem and $k \cdot p$ theory

$k \cdot p$ theory is a perturbation method, whereby if we know the exact energy levels at one point in the Brillouin zone (say $k = 0$, the Γ point) then we can use perturbation theory to calculate the band structure near that k value. We use $k \cdot p$ theory in Chapters 4 and 5 to explain various aspects of the electronic structure of semiconductors. A general introduction to first and second order perturbation theory is given in Appendix C.

The Hamiltonian, H_0, in a periodic solid is given by

$$H_0 = -\frac{\hbar^2}{2m}\nabla^2 + V(r) \tag{E.1}$$

with $V(r+R) = V(r)$, as discussed in Chapter 3. We also saw in Section 3.2 how the eigenstates, $\psi_{nk}(r)$, can be written using Bloch's theorem as the product of a plane wave, $e^{ik \cdot r}$, times a periodic function, $u_{nk}(r)$, with associated energy levels, E_{nk}. For a particular value of k, say k_0, Schrödinger's equation may be written as

$$H_0\psi_{nk_0}(r) = \left(-\frac{\hbar^2}{2m}\nabla^2 + V(r)\right)(e^{ik_0 \cdot r}u_{nk_0}(r)) = E_{nk_0}(e^{ik_0 \cdot r}u_{nk_0}(r)) \tag{E.2}$$

We presume that we know the allowed energy levels E_{nk_0} at k_0 and now wish to find the energy levels, E_{nk}, at a wavevector k close to k_0, where

$$\left(-\frac{\hbar^2}{2m}\nabla^2 + V(r)\right)(e^{ik \cdot r}u_{nk}(r)) = E_{nk}(e^{ik \cdot r}u_{nk}(r)) \tag{E.3}$$

To emphasise that we are interested in values of k close to k_0, we may rewrite eq. (E.3) as

$$\left(-\frac{\hbar^2}{2m}\nabla^2 + V(r)\right)e^{i(k-k_0) \cdot r}(e^{ik_0 \cdot r}u_{nk}(r)) = E_{nk}e^{i(k-k_0) \cdot r}(e^{ik_0 \cdot r}u_{nk}(r)) \tag{E.4}$$

We would normally describe Schrödinger's equation as a second order differential equation acting on the full wavefunction, $\psi_{nk}(r)$. We can also, however, view eq. (E.4) as a second order differential equation involving the unknown function $\exp(ik_0 \cdot r)u_{nk}(r)$. If we multiply both sides of eq. (E.4) from the left by $\exp[-i(k-k_0)\cdot r]$, we obtain a modified differential equation from which to determine E_{nk}:

$$[e^{-i(k-k_0)\cdot r}H_0e^{i(k-k_0)\cdot r}](e^{ik_0\cdot r}u_{nk}(r)) = [e^{-i(k-k_0)\cdot r}E_{nk}e^{i(k-k_0)\cdot r}](e^{ik_0\cdot r}u_{nk}(r))$$

$$= E_{nk}(e^{ik_0\cdot r}u_{nk}(r)) \qquad (E.5)$$

Between eqs (E.3) and (E.5), we have transformed from a k-dependent wavefunction, ψ_{nk}, to a k-dependent Hamiltonian, which we write as H_q, where $q = k - k_0$. Equation (E.5) can be re-written as

$$H_q\phi_{nk}(r) = e^{-iq\cdot r}\left(-\frac{\hbar^2}{2m}\nabla^2 + V(r)\right)e^{iq\cdot r}\phi_{nk}(r) \qquad (E.6)$$

where $\phi_{nk}(r) = \exp(ik_0 \cdot r)u_{nk}(r)$. We now expand the term $\nabla^2 e^{iq\cdot r}\phi_{nk}(r)$ to obtain

$$H_q\phi_{nk}(r) = \left(-\frac{\hbar^2}{2m}\nabla^2 + \frac{\hbar^2}{m}q\cdot\frac{1}{i}\nabla + \frac{\hbar^2q^2}{2m} + V(r)\right)\phi_{nk}(r)$$

$$= \left[H_0 + \frac{\hbar}{m}q\cdot p + \frac{\hbar^2q^2}{2m}\right]\phi_{nk}(r) \qquad (E.7)$$

where we have used eq. (E.1), and replaced $\hbar/i\nabla$ by the momentum operator, p, introduced in Chapter 1.

Equation (E.7) forms the basis of the $k \cdot p$ method. It reduces to the standard form of Schrödinger's equation when $q = 0$, at the point k_0. For many applications, we choose $k_0 = 0$, the Γ point, where we generally know or can estimate the values of all the relevant zone centre energies, E_{n0}. We can then view

$$H' = \frac{\hbar}{m}q\cdot p + \frac{\hbar^2q^2}{2m} \qquad (E.8)$$

as a perturbation to the zone centre Hamiltonian, H_0, and use second order perturbation theory to calculate the variation of the energy levels E_{nk} with wavevector $k(=q)$ close to the Γ point.

For the case of a singly degenerate band, substituting eq. (E.8) into eq. (C.23) gives the energy of the nth band in the neighbourhood of $k = 0$ as

$$E_{nk} = E_{n0} + \frac{\hbar}{m}k\cdot p_{nn} + \frac{\hbar^2k^2}{2m} + \frac{\hbar^2}{m^2}\sum_{n'\neq n}\frac{|k\cdot p_{nn'}|^2}{E_{n0}-E_{n'0}} \qquad (E.9)$$

where $p_{nn'}$ is the momentum matrix element between the nth and n'th zone centre states, $u_{n0}(r)$ and $u_{n'0}(r)$

$$p_{nn'} = \int_v d^3r \, u_{n0}^*(r) p u_{n'0}(r) = \langle u_{n0}|p|u_{n'0}\rangle \tag{E.10}$$

with the integration in (E.10) taking place over a unit cell of the crystal structure.

Hence, we can write the energy at some general wavevector k in terms of the known energies at $k = 0$, and the interactions between the zone centre states through the momentum matrix elements $p_{nn'}$. The term $(\hbar/m)k \cdot p_{nn}$ is linear in k, while the other two terms in eq. (E.9) are quadratic in k. Kane (1966) shows that the linear term is by symmetry identically equal to zero in diamond structures and estimates that its effects are negligibly small and can generally be ignored in III–V semiconductors.

We often describe the band dispersion near the zone centre as if the carriers have an effective mass, m^*, with

$$E_{nk} = E_{n0} + \frac{\hbar^2 k^2}{2m^*} \tag{E.11}$$

where $1/m^*$ may depend on direction. Comparing eqs (E.9) and (E.11) we find the effective mass m_i^* is given using $k \cdot p$ theory by

$$\frac{1}{m_i^*} = \frac{1}{m^*} + \frac{2}{m^2} \sum_{n' \neq n} \frac{|i \cdot p_{nn'}|^2}{E_{n0} - E_{n'0}} \tag{E.12}$$

where i is a unit vector in the direction of the ith principal axis.

For practical applications, we need to include the effect of spin–orbit interaction, particularly at the valence band maximum. We should also take into account band degeneracies, such as that between the heavy-hole and light-hole bands at the valence band maximum, where we strictly need to use degenerate perturbation theory. We include these effects implicitly in the discussion in Chapter 4. A more detailed derivation and description of the application of $k \cdot p$ theory to semiconductors may be found for instance in Kane (1963).

Reference

Kane, E. O. (1963) *Semiconductors and Semimetals*, Vol. 1, Ch. 3, ed. R. K. Willardson and A. C. Beer, Academic Press, New York.

Outline solutions to problems

Chapter 1

1.1 For odd states, the wavefunction is given by $\psi(x) = A\sin(kx)$ inside the well; $D\,e^{-\kappa x}$ for $x > a/2$. Adapt eq. (1.49) to get the answer.

1.2 First normalise by integrating $a_n^2 \sin^2(n\pi x/L)$ from 0 to L, giving $a_n^2 L/2$, so that $a_n = (2/L)^{1/2}$. Then integrate $\psi_n^*\psi_m$ using $\sin(n\pi x/L)\sin(m\pi x/L) = \frac{1}{2}(\cos[(n-m)\pi x/L] - \cos[(n+m)\pi x/L]$ to show integral zero when $n \neq m$.

1.3 Because ψ_1 and u are orthogonal, we can write

$$\langle E \rangle = \frac{\int \psi_1^* H \psi_1 \, dx + \epsilon^2 \int u^* H u \, dx}{\int \psi_1^* \psi_1 \, dx + \epsilon^2 \int u^* u \, dx}$$

Using the normalisation of the functions, we find $\langle E \rangle = E_1 + \epsilon^2 [\int u^* H u \, dx - E_1] + O(\epsilon^4)$.

1.4 (a) We find $Hf(x) = (\hbar^2/m)n(n-1)x^{n-2}$. Evaluating eq. (1.37), we find

$$\langle E \rangle = \frac{1}{L^2}\frac{\hbar^2}{m}\frac{(n+1)(2n+1)}{2n-1}$$

The estimated value of $\langle E \rangle$ increases rapidly with n, as the shape of the function starts to deviate significantly from the true wavefunction.

(b) $g(x)$ is the simplest polynomial which is odd about $x = 0$, and zero at $\pm L/2$. We again evaluate eq. (1.37) to find $\langle E_2 \rangle = 21\hbar^2/mL^2$, compared to the true value $2\pi^2\hbar^2/mL^2$.

(c) We need an even polynomial, equal to zero at $\pm L/2$, with two other zeros between $\pm L/2$; so of the form $f_2(x) = A_1 - B_1 x^2 + C_1 x^4$. Exercise: find A, B, C that satisfies above and also generates a quartic polynomial orthogonal to the ground state variational function $(L/2)^2 - x^2$. Likewise next odd state of form $g_2(x) = A_2 x - B_2 x^3 + C_2 x^5$, and orthogonal to $g(x)$.

1.5 We again need to evaluate eq. (1.37). We end up in this case with two terms in the estimated $\langle E \rangle$, one, $\langle T \rangle$, associated with the kinetic energy term $(-\hbar^2/2m)\mathrm{d}^2/\mathrm{d}x^2$ in the Hamitonian, and the other $\langle V \rangle$ from the potential energy term, $\frac{1}{2}kx^2$. We find

$$\langle E \rangle = \langle T \rangle + \langle V \rangle = \frac{\hbar^2\alpha}{2m} + \frac{k}{8\alpha}$$

Differentiating with respect to α, this is minimised when $\alpha = (km/\hbar^2)^{1/2}$, so that $\langle E \rangle = \frac{1}{2}\hbar(k/m)^{1/2}$. This is the correct ground state energy. Likewise, using the trial function $g(x)$, which is odd about $x = 0$, gives the correct first excited state energy, $\frac{3}{2}\hbar(k/m)^{1/2}$.

Chapter 2

2.1 Follow the same procedure as in eqs (2.1)–(2.9), but replace eq. (2.2) by $\psi(x) = A(e^{\kappa x} - e^{-\kappa x})$ in the central region ($|x| < b/2$) to derive the solution for states of odd parity.

2.2 Replace $\tanh(\kappa b/2)$ by $1 - 2e^{-\kappa b/2}\operatorname{sech}(\kappa b/2)$ in eq. (2.9b); keep terms involving 1 on the LHS and take the two terms involving $2e^{-\kappa b/2}\operatorname{sech}(\kappa b/2)$ to the RHS. Divide both sides by 2. LHS is now the same as eq. (1.54), so can be replaced by the product of two terms in eq. (1.53); dividing both sides of equation by $k\cos(ka/2)+\kappa\sin(ka/2)$ gives required eq. (2.38).

For the second part of the question, expand $f(E) = f(E_0)+ (E - E_0)f'(E_0)$. But we know $f(E_0) = 0$. Also $\operatorname{sech}(\kappa b/2) \sim e^{-\kappa b/2}$ for large b, so we can therefore say $(E - E_0)f'(E_0) \sim e^{-\kappa b}$, which can then be re-arranged to give the required result.

2.3 Replace $\coth(\kappa b/2)$ by $1 + 2e^{-\kappa b/2}\operatorname{cosech}(\kappa b/2)$ in problem 2.1 and then follow same procedure as for problem 2.2 to get $E_{\mathrm{ex}}(b) = E_0 + Ce^{-\kappa b}$.

2.4 By setting $D = A\cos(ka/2)\exp(\kappa a/2)$, and integrating $|\psi(x)|^2$ we find

$$\frac{A^2}{2}\left[a + \frac{\sin(ka)}{k} + \frac{1 + \cos(ka)}{\kappa}\right] = 1$$

Note that this reduces correctly to $A^2 = 2/a$ in an infinite square well, where $\kappa = \infty$, and $\sin(ka) = 0$.

2.5 We can show that $\mathbf{I} = E_0 - V_0\int_b^{b+a} D^2 e^{-2\kappa x}\,\mathrm{d}x \sim E_0 - C_{\mathrm{I}}e^{-2\kappa b}$. Likewise, $\mathbf{IV}$ varies as $C_{\mathrm{IV}}e^{-\kappa b}$. Finally, we can show $\mathbf{II}$ also contains terms that vary as $C_{\mathrm{II}}e^{-\kappa b}$, which dominate at large b. When we solve the resulting 2×2 determinant, the terms in **II** have the dominant effect in determining the splitting for large b, which then varies for large b as $2C_{II}e^{-\kappa b}$.

2.6 I have never attempted to prove this, but presume it is true!

2.7 We require $\partial \phi_h / \partial \phi = 0$, and $\partial \phi_h / \partial \theta = 0$. Evaluating $\partial \phi_h / \partial \phi = 0$, we find $\cos \phi = \sin \phi$, and $\phi = \pi / 4$. If we now evaluate $\partial \phi_h / \partial \theta = 0$, we find $\tan \theta = -\sqrt{2}$, for which we can deduce $\cos \theta = 1/\sqrt{3}$. We must also show by considering second derivatives that this solution gives a maximum pointing along the (111) direction.

2.8 The three solutions are

$$\psi_1 = \frac{1}{\sqrt{3}} \left(\phi_s + \sqrt{2}\, \phi_x \right);$$

$$\psi_2 = \frac{1}{\sqrt{3}} \left(\phi_s - \frac{1}{\sqrt{2}}\, \phi_x + \frac{\sqrt{3}}{\sqrt{2}}\phi_y \right);$$

$$\psi_3 = \frac{1}{\sqrt{3}} \left(\phi_s - \frac{1}{\sqrt{2}}\, \phi_x - \frac{\sqrt{3}}{\sqrt{2}}\phi_y \right).$$

2.9 Multiply $H\psi_n = E_n\psi_n$ on the left by $e^{-i(2\pi jn/N)}\phi_j$ and integrate to find $E_a + U(e^{i2\pi n/N} + e^{-i2\pi n/N}) = E_n[1 + S(e^{i2\pi n/N} + e^{-i2\pi n/N})]$. If $S \sim 0$, this simplifies to the required result. For an N-membered ring, add energies of N lowest levels (remembering double degeneracy), and then divide by N to get average binding energy per atom. Binding energy is greatest for six-membered ring. This is also the case where bond angle is 120°, thereby also maximising sp^2 contribution to bonding.

2.10 $E_g = h\upsilon$; $\lambda \upsilon = c$. Eliminating υ, we find $E_g\lambda = hc = 1.986 \times 10^{-25}$ J m. Convert energy units to eV and length to μm to give required result. $dE_g/dx = 0$ when $x = 0.336$; $E_g = 0.104$ eV; $\lambda \approx 12$ μm.

Chapter 3

3.1 First electron goes into state at $q = 0$; then start to fill states with increasing $|k|$, until $|k| = \pi/L$, when two electrons per atom. We therefore need to evaluate

$$E_{bs} = 2V\frac{L}{\pi} \int_{-\theta}^{\theta} \cos(qL)\, dq$$

where $\theta = y\pi/2L$. This gives $E_{bs} = 4V/\pi \sin(y\pi/2)(=2Vy$ at small y; $4V/\pi$ when $y = 1$).

3.2 Let the wavefunction of state with wavevector q, $\psi_q = \alpha\phi_0 + \beta e^{iqL}\phi_1$ in the zeroth unit cell. We evaluate $\int \phi_{0,1}^* H \psi_q$ to get two linear equations, the first of which is

$$\alpha E_s + \beta e^{iqL}(V + \Delta V) + \beta e^{-iqL}(V - \Delta V) = \alpha E_{sq}$$

Solving these equations gives the required band structure, E_{sq}. Because unit cell size is now doubled, Brillouin zone edges are now at $\pm\pi/2L$, where we find $E_{sq} = E_s \pm 2\Delta V$. The band-structure energy gained per atom, E_{bs}, due to the distortion is given by the solution of the integral

$$E_{bs} = \frac{4}{\pi}\int_0^{\pi/2}\left(\sqrt{V^2\cos^2\theta + (\Delta V)^2\sin^2\theta} - |V|\cos\theta\right)d\theta$$

This integral does not have a simple solution: its value $E_{bs} = 4|V|/\pi[E((1-a^2)^{1/2}) - E(1)]$, where $a = \Delta V/V$ and E is an elliptical integral. With considerable approximation, it can then be shown that $E_{bs} \sim (\Delta V)^2/|V|$.

3.3 We choose the nth NFE wavefunction at $q = \pi/L$ as $\psi = \alpha e^{i(2n-1)\pi/L} + \beta e^{-i(2n-1)\pi/L}$. Follow the same analysis as in eqs (3.32)–(3.39) to get required results. At the zone centre, replace $2n - 1$ by $2n$ in analysis.

3.4 Need to show $\alpha = \beta$ for the upper state, and $\alpha = -\beta$ for the lower state, respectively. Because the wavefunction of the lower state has a node near $x = 0$, in region of Kronig–Penney (K–P) barrier, it is less perturbed by the K–P potential, and so provides a better energy estimate to larger b than the upper state.

3.5 We choose $\psi(x) = \alpha\sin(\pi x/L) + \beta\sin(3\pi x/L)$ as our trial function for the lowest state, which we know is odd about $x = 0$. We then follow the analysis of eqs (3.32)–(3.39), and choose the lower of the two solutions obtained as our improved estimate for the lower state. We then replace the sine by cosine functions; same analysis gives the improved estimate for the upper level.

3.6 $\lambda = 2\pi/(k_x^2 + k_y^2)^{1/2} = 2\pi/k$. $v = k/\omega$. The wave propagates along the direction $(\cos\theta, \sin\theta)$. For $t = 0$, $\psi = 1$ when $8\pi/L(x+y) = 2n\pi$, that is, $x + y = nL/4$. We can draw solutions for this in the given square for $0 \leq n \leq 8$. We see that the repeat distance in the $x(y)$-direction is just $L/4 (= 2\pi/k_{x(y)})$, while along the propagation direction $\lambda = \sqrt{2}L/8$. Show also that at $t = \pi/\omega$, the lines $\psi = 1$ have moved forward by half a wavelength.

3.7 We need to show that we can generate the six second neighbours, for example, $(0,0,a) = a_1 + a_2 - a_3$; also the other nine first neighbours, for example, $(0, -a/2, a/2) = a_2 - a_3$. Any other lattice point can then readily be generated from a sum of first and second neighbours.

3.8   We can show in general that if $R = n_1 a_1 + n_2 a_2 + n_3 a_3$ and $G = m_1 b_1 + m_2 b_2 + m_3 b_3$, then $R \cdot G = 2\pi(n_1 m_1 + n_2 m_2 + n_3 m_3)$, so G is a reciprocal lattice vector. Also no reciprocal basis vectors are missing, for example, if $(n_1, n_2, n_3) = (1, 0, 0)$ then $R \cdot G = 2\pi m_1$, and as m_1 is any integer, all values of $2\pi n$ are included.

3.9   Show that $b_1 = (2\pi/a)(-1, 1, 1)$; $b_2 = (2\pi/a)(1, -1, 1)$; $b_3 = (2\pi/a)(1, 1, -1)$, and that these are the basis vectors of a BCC lattice with the eight first neighbours at $(2\pi/a)(\pm1, \pm1, \pm1)$ and six second neighbours at $(4\pi/a)(1, 0, 0)$ etc. Working in the opposite direction one can also show reciprocal lattice of a BCC lattice is an FCC lattice.

3.10 We want to find b_1 and b_2 in the x–y plane, so that the reciprocal lattice is given by $G = m_1 b_1 + m_2 b_2$. Most easily done by introducing $a_3 = (0, 0, a)$. We find $b_1 = (2\pi/a)(1, -1/\sqrt{3})$; $b_2 = (2\pi/a)(0, -2/\sqrt{3})$, which it can be shown are also the basis vectors of a triangular lattice.

3.11 Consider, for example, the atom at the origin, which has six neighbours, with coordinates $c_1, \ldots, c_6$ at $\pm a_1$, $\pm a_2$, $\pm(a_1 - a_2)$. Then evaluate $E_{sk} = E_s + \sum_j V \exp(i k \cdot c_j)$ to get the required answer. For the lower band edge at $k = 0$, wavefunction is in phase on all sites, so we get $E_{sL} = E_s - 6|V|$. By contrast, one cannot get an antibonding state where phase is of opposite sign on all neighbours – the best that can be managed in a periodic wavefunction is to have effectively four neighbours of opposite sign and two of same sign, giving $E_{sU} = E_s + 2|V|$.

Chapter 4

4.1   Invert (4.9a) to determine values of E_p. Note that the values do not vary strongly between the different materials.

4.2   Substituting E_p into eq. (4.9b) we calculate, for example, $m_{lr}^* = 0.11$ (0.056) for GaAs (GaSb), compared to 0.082 (0.05) in Table 4.1.

4.3   We calculate, for example, for InP that $E_1^{imp} = 12\,\mathrm{meV}$ and $a^* = 64\,\text{Å}$, so that $N_d \sim (1/64)^3\,\text{Å}^{-3} \sim 4 \times 10^{18}\,\mathrm{cm}^{-3}$.

4.4   Follow analysis in Appendix E to get solution.

4.5   We evaluate the integral

$$P_{LUn} = \frac{2}{L} \int_0^L \sin\left(\frac{n\pi x}{L}\right) \left(-i\hbar\frac{d}{dx}\right) \cos\left(\frac{n\pi x}{L}\right) dx$$

to get the required result. This partly explains the relative constancy of E_p in problem 4.1, as the bond length does not change strongly between the materials listed.

4.6 We find for wavevectors q close to π/L that eq. (3.38) and the $\mathbf{k} \cdot \mathbf{p}$ method give the same result, with

$$E = \frac{b}{L}V_0 + \frac{\hbar^2}{2m}\left[\left(\frac{\pi}{L}\right)^2 + \left(q - \frac{\pi}{L}\right)^2\right]$$

$$\pm \left[\frac{V_0}{\pi}\sin\frac{\pi b}{L} + \frac{\hbar^4\pi^3}{2m^2L^2V_0\sin(\pi b/L)}\left(q - \frac{\pi}{L}\right)^2\right]$$

4.7 We first use the $\mathbf{k} \cdot \mathbf{p}$ method to show that first order wavefunction of the upper state varies with wavevector k as $\psi_{Uk}(x) = (2/L)^{1/2}[\cos(2\pi x/L) + iAk\sin(2\pi x/L)]$, where $A = 2\pi^2\hbar^2/[mLV_0\sin(2\pi b/L)]$, with a related expression for $\psi_{Lk}(x)$. We then use eq. (1.37) to estimate the average potential energy at wavevector k in the upper and lower bands, $\langle U_{U/L}(k)\rangle$ to find

$$\langle U_{U/L}(k)\rangle = \frac{V_0 b}{L} \pm \frac{V_0}{2\pi}\sin(2\pi b/L)[1 - 2|A|^2k^2]$$

This shows that the potential energy, $\langle U_U(k)\rangle$ decreases with increasing k in the upper band, and increases with k in the lower band. As total energy $\langle E\rangle = \langle T\rangle + \langle U\rangle$, we then find that the kinetic energy $\langle T\rangle$ must increase with k in the upper band, and *decrease* with increasing k in the lower band. This result can be used to explain why the lower band has a negative effective mass at the zone centre.

4.8 The average acceleration $\langle a\rangle$ is found by integrating over all wavevectors k between $-\pi/L$ and π/L. We find

$$\int_{-\pi/L}^{\pi/L} a\,dk = \frac{1}{\hbar^2}\left[\frac{dE}{dk}F\right]_{-\pi/L}^{\pi/L}$$

Result then follows as $dE/dk = 0$ at $\pm\pi/L$.

Chapter 5

5.1 We require $\hbar^2\pi^2/(2mL^2)(m_{LH}^{*-1} - m_{HH}^{*-1}) = 40\,\text{meV}$. We find $L = 94\,\text{Å}$.

5.2 We calculate $\Delta E_c = 0.1742\,\text{eV}$ and $\Delta E_v = 0.0938\,\text{eV}$. With $\hbar^2 k^2/2m^* = 25\,\text{meV}$, we get, for example, $k = 0.021\,\text{Å}^{-1}$ for $m_e^* = 0.067$, and $\kappa/k = 2.442$. Substituting in equation, we find $L_e = 113\,\text{Å}$.

5.3 The maximum possible value of k^2 occurs for a confined state when $\kappa = 0$ and $\hbar^2 k^2/2m^* = \Delta E_c$ (eq. A). To get a solution of $-k\cot(ka) = \kappa$, we require that the LHS is positive, and $(ka)^2 > (\pi/2)^2$ (eq. B). Combining eqs (A) and (B) gives the required result. For the second bound state, we require $ka > 3\pi/2$, and hence $\Delta E_c a^2 > 9\pi^2\hbar^2/8m^*$.

5.4 We require $a^2 > \pi^2\hbar^2/(8m^*\Delta E_c)$, that is, $a > 21.7$ Å.

5.5 This is a matrix inversion problem: either calculate directly the inverse of the matrix R or show that the matrix product $RG = I$, the identity matrix.

5.6 We need to evaluate the integral

$$n = \int_{E_c}^{\infty} dE \, \frac{m_c^*}{\pi\hbar^2} f_c(E)$$

with $f_c(E)$ given by eq. (5.37) to get the required result. Then re-arrange $\pi\hbar^2 n/(m_c^* kT) = \ln[\exp\{(F_c - E_c)kT\} + 1]$ to determine how F_c varies with n. Similarly for p and F_v.

5.7 First show that we can rewrite eq. (5.37) as $f_c = 1 - [\exp\{(F_c - E_c)/kT\} + 1]^{-1}$. The term in square brackets here can be shown to be equal to the second term on the RHS of the equation for $g_{\max}$, while the last term on the RHS equals f_v. We calculate the transparency carrier density by solving the RHS of the equation for $g_{\max} = 0$. The carrier density n for a given quasi-Fermi energy, is found by integrating over the density of states, $n = \int_0^{\infty} dE \, g(E) f_c(E)$ where we have set the band edge energy to zero. As $g(E)$ varies as $E^{(D-2)/2}$ near the band edge, we can show for $F_c = 0$ (i.e. at the band edge) that n varies as $(kT)^{D/2}$. This is the condition for transparency when $m_c^* = m_v^*$. It can be shown that the decreased temperature dependence of n_0 when the dimensionality D is reduced leads also to a decrease in the temperature dependence of the transparency (and hence threshold) current density.

Chapter 6

6.1 Setting $J = \frac{1}{2}$ in eq. (6.35), and letting $x = \frac{1}{2}g\mu_B B/kT$, we get

$$\langle \mu \rangle = \frac{1}{2}g\mu_B \frac{e^x - e^{-x}}{e^x + e^{-x}}$$

as required. Using $\tanh(x) \sim x$ for small x, and with N ions per unit volume, we have $M = \frac{1}{2}Ng\mu_B x$, and $\chi = N\mu_0(g\mu_B)^2/4kT$.

6.2 Let $y = g\mu_B B/kT$ in eq. (6.35). The denominator is given by the geometric series $S(y) = e^{-Jy}[1 + e^y + \cdots + e^{2Jy}] = \sinh\{(J + \frac{1}{2})y\}/\sinh(\frac{1}{2}y)$. We can show the numerator $\sum m e^{my} = dS/dy$. Dividing gives the required result. As $J \to \infty$, $(2J+1)/2J \to 1$, and $\coth(x/2J) \to 2J/x$, giving the classical result. By expanding $\cosh$ and $\sinh$ to order x^2, show $\coth(x) = 1/x + \frac{1}{3}x$. The two terms in $1/x$ then cancel in $B_J(x)$, giving $B_J(x) \sim (J+1)x/(3J)$, which when substituted into eq. (6.36) gives the required results.

6.3 We first evaluate $\langle R_i^2 \rangle = \int_0^\infty r^2 |\psi(r)|^2 4\pi r^2 \, dr = 3a_0^2$. Substituting into eq. (6.17) we find $\chi_{\text{dia}} = -5 \times 10^{-35}N$, where N is the number of H atoms per m^3. The paramagnetic susceptibility $\chi_{\text{para}} = N\mu_0\mu_B^2/kT$. We then find $\chi_{\text{para}} = \chi_{\text{dia}}$ when $T \sim 1.5 \times 10^5$ K, demonstrating that a gas of H atoms would be paramagnetic at room temperature.

6.4 For Cr^{3+}, with three d electrons, $L = \sum m = 2 + 1 + 0 = 3$, $S = \sum s = 3/2$, and $J - L - S = 3/2$. Calculate g, and evaluate the two different expressions to show $2(1.5 \times 2.5)^{1/2}$ is the better fit.

6.5 $l = 3$ for f-shell. Shell over half-full, with 8th and 9th electrons have $m = 3, 2$. So $L = \sum m = 5$. $S = 5/2$. Hund's 3rd rule gives $J = L + S = 15/2$. Multiplicity $= 2S + 1 = 6$, so notation for ground state is $^6H_{15/2}$. Predicted $g = 1.33$, so $p = 1.33(7.5 \times 8.5)^{1/2} = 10.645$.

6.6 (a) Inequality follows from showing $\nabla \cdot \mathbf{F} = (V\chi/\mu_0)\nabla^2 B^2$.
   (b) $\nabla \cdot \mathbf{B} = 0 \Rightarrow \partial B_x/\partial x = -\partial B_y/\partial y - \partial B_z/\partial z$ (eq. (1)). $\nabla \times \mathbf{B} = 0 \Rightarrow \partial B_z/\partial x = \partial B_x/\partial z$ (eq. (2)), and two equivalent equations. Take the derivative of eq. (1) with respect to x, and use eq. (2) on the right-hand side to show $\nabla^2 B_x = 0$.
   (c) We can show by double differentiation that $\nabla^2 B_x^2 = 2B_x\nabla^2 B_x + 2|\nabla B_x|^2 = 2|\nabla B_x|^2 \geq 0$. As $\chi > 0$ for a paramagnet, must then always have $\nabla \cdot \mathbf{F} > 0$ for a paramagnet.

6.7 We have $B_x(x, 0, z) = B_x(0, 0, z) + x\partial B_x/\partial x(0, 0, z) + \cdots$, with $B_x(0, 0, z) = 0$ from symmetry. From rotational symmetry, $\partial B_x/\partial x(0, 0, z) = \partial B_y/\partial y(0, 0, z)$ and these $= -\frac{1}{2}\partial B_z/\partial z(0, 0, z)$ (eq. (3)) using $\nabla \cdot \mathbf{B} = 0$. For the z component, we use $B_z(x, 0, z) = B_z(0, 0, z) + x\partial B_z/\partial x(0, 0, z) + \frac{1}{2}x^2\partial^2 B_z/\partial x^2(0, 0, z)$. The second term on the right is zero from symmetry. Taking the derivative of eq. (3) with respect to z, and then using eq. (2) from problem 6.6, we can show that $\partial^2 B_z/\partial x^2 = -\frac{1}{2}\partial^2 B_z/\partial z^2$, giving required result.

6.8 The stability conditions require $\partial^2 B^2(0, 0, z)/\partial x^2 > 0$, and $\partial^2 B^2(0, 0, z)/\partial z^2 > 0$. From problem 6.7, we find retaining terms to x^2 that

$$B^2(x, 0, z) = \frac{1}{4}[B^{(1)}(0, 0, z)]^2 x^2 + [B(0, 0, z)]^2 - \frac{1}{2}B(0, 0, z)B^{(2)}(0, 0, z)x^2$$

Evaluating the two second derivatives, the first requires $[B^{(1)}(0, 0, z)]^2 - 2B(0, 0, z)B^{(2)}(0, 0, z) > 0$, that is, $z < (2/5)^{1/2}a$, while the second requires $[B^{(1)}(0, 0, z)]^2 + B(0, 0, z)B^{(2)}(0, 0, z) > 0$, which gives $z > a/\sqrt{7}$.

Chapter 7

7.1 The magnetisation, $M = M_s \tanh(x)$ (eq. (A)), where $x = \mu_0 \mu_B \lambda M/kT$ (eq. (B)) and $M_s = N\mu_B$. At small x, eq. (A) becomes $M = M_s x$ (eq. (C)). At $T = T_c$, eqs (B) and (C) have the same slope, so that $M_s = kT_c/\mu_0 \mu_B \lambda$, and $T_c/\lambda = \mu_0 \mu_B M_s/k$. We can also show $x = MT_c/M_s T$. Substituting in eq. (A), we then find (using $y = \tanh(x) \Rightarrow x = \frac{1}{2} \ln[(1+y)/(1-y)]$ that $M/M_s = 0.8$ when $0.8 = \tanh(0.8T_c/T)$, or $T = 0.728T_c$. Likewise $M/M_s = 0.5$ when $T = 0.910T_c$.

7.2 At $T = 0$, $x = \infty$, and $M = M_s$. From Section (7.3), we have

$$M = \frac{C}{T - T_c}\frac{B}{\mu_0} \qquad \text{(D)}$$

Substituting for $C = T_c/\lambda = \mu_0 \mu_B M_s/k$ in eq. (D), we find that the ratio of M at $300\,K$ to M at $0\,K$ is given by $\mu_B B/k(T - T_c) \approx 6 \times 10^{-3}\,meV/12.5\,meV \sim 5 \times 10^{-4}$.

7.3 α_i is the direction cosine with respect to the ith axis. For (100), $\alpha_1 = 1$; $\alpha_{2,3} = 0$. For (110), $\alpha_{1,2} = 1/\sqrt{2}$; $\alpha_3 = 0$. For (111), $\alpha_{1,2,3} = 1/\sqrt{3}$. Substitute these values in eq. (7.36a) to find $W = K_1/4$ and $K_1/3 + K_2/27$, respectively.

7.4 For small T (large x), the straight line (eq. (7.17)) cuts the tanh curve at $x = MT_c/M_s T \sim T_c/T$ as $M \sim M_s = N\mu_B$. At large x, $M = M_s \tanh(x) \approx M_s(1 - 2e^{-2x}) = M_s(1 - 2\exp(-2T_c/T))$. Near the Curie temperature, we have for small x that $M = M_s \tanh(x) \approx M_s(x - x^3/3)$ (eq.(E)). Substituting that $x = MT_c/M_s T$ in eq. (E), and re-arranging, we can show that in mean field theory, $M = \sqrt{3}M_s(T/T_c)(1 - T/T_c)^{1/2} \to \sqrt{3}M_s(1 - T/T_c)^{1/2}$ as $T \to T_c$.

7.5 In Langevin theory, $M \to M_s(1 - 1/x)$ as $x \to \infty$. We also require that $x = MT_c/M_s T$. Following the same process as in problem 7.4, we then derive $M = M_s(1 - T/T_c)$. The linear variation here is closer to the experimentally observed behaviour because low energy excitations are possible in the classical model, with average spin deflection $\langle \theta \rangle$ from the field direction $\to 0$ smoothly as $T \to 0$.

7.6 Because Fe^{3+} moments cancel, we have net magnetic moment per unit cell is due to eight Ni^{2+}. When orbital angular momentum is quenched, $J = S = 1$; $g = 2$ for Ni^{2+}, with the net moment per Ni ion $= 2\mu_B$, giving total moment of $16\mu_B$. When two non-magnetic Zn displace two Fe^{3+} to the B sites, the net moment is now due to six Ni^{2+} and four unbalanced Fe^{3+} ions. From fig. 6.6, each Fe^{3+} has moment of $5\,\mu_B$, giving a total moment of $32\,\mu_B$.

7.7 We have $M_A = \frac{1}{2}(C_A/T)(H_0 - \lambda M_B)$, and an equivalent expression for M_B. We can solve these two linear equations to see that $M = M_A + M_B$ varies with H_0 as

$$M = \frac{\lambda T(C_A + C_B) - \lambda^2 C_A C_B}{2T^2 - \lambda^2 C_A C_B/2} \frac{H_0}{\lambda}$$

The susceptibility, $\chi \to \infty$ when the bottom line is zero, giving $T_c = \lambda(C_A C_B)^{1/2}/2$. It can be shown that the temperature dependence of the inverse ferrimagnetic susceptibility is then markedly different to that of a ferromagnet, for which $\chi^{-1} = (T + T_c)/C$.

Chapter 8

8.1 Just because $\nabla \times E = 0$ does not imply $E = 0$. You are probably familiar with the idea that an oscillating current in an aerial generates an electromagnetic wave, with the time-dependent E-field polarised parallel to the aerial axis. In the same way, the decay of the current in the solenoid gives rise to the E-field which accelerates the electron as derived in Section 8.5.

8.2 Evaluate one (e.g. x) component of each side of the equation. We first evaluate $\nabla \times j = i(\partial j_z/\partial y - \partial j_y/\partial z) + j(\partial j_x/\partial z - \partial j_z/\partial x) + k(\partial j_y/\partial x - \partial j_x/\partial y)$, where i, j and k are unit vectors along the co-ordinate axes. The x-component of $\nabla \times (\nabla \times j)$ is then equal to $\partial/\partial y(\partial j_y/\partial x - \partial j_x/\partial y) - \partial/\partial z(\partial j_x/\partial z - \partial j_z/\partial x)$. The x-component of $\nabla(\nabla \cdot j) = \partial/\partial x(\partial j_x/\partial x + \partial j_y/\partial y + \partial j_z/\partial z)$, while $\nabla^2 j_x = \partial^2 j_x/\partial x^2 + \partial^2 j_x/\partial y^2 + \partial^2 j_x/\partial z^2$. Simplifying both sides of the equation, we verify the expression. The conservation of charge relates the current density j to the free charge density ρ at each point by $\nabla \cdot j = -\partial \rho/\partial t$. As $\partial \rho/\partial t = 0$ in steady state, we therefore require $\nabla \times (\nabla \times j) = -\nabla^2 j$. Taking the curl of both sides of eq. (8.18) then gives $-\nabla^2 j = -n_s e^2/m_e(\nabla \times B)$, which using eq. (8.19a) equals $-(\mu_0 n_s e^2/m_e)j$, identical to eq. (8.22) for B, so that when solved the solution must be of similar form. As $j = \mu_0 \nabla \times B$, we find by taking the curl of $B = B_0 k \exp(-x/\lambda_L)$ that $\mu_0 j_0 \lambda_L = B_0$, as required. Finally, we find the magnetic flux penetrating per unit length by evaluating $\int_0^\infty B_0 \exp(-x/\lambda_L) \, dx$ to get the required result.

8.3 From eq. (8.5), $H_c(T) = H_c(0)[1 - (T/T_c)^2]$, and so $\partial H_c/\partial T = -2H_c(0)T/T_c^2$. Substituting into $\Delta S = -\mu_0 H_c \partial H_c/\partial T$, we obtain $\Delta S(T) = 2\mu_0 H_c(0)^2/T_c[(T/T_c) - (T/T_c)^3]$. Taking the derivative of ΔS with respect to T, we find the derivative equals zero, and ΔS is maximised when $1 - 3(T/T_c)^2 = 0$, that is, when $T = T_c/\sqrt{3}$, as required.

8.4 We have $\Delta S(T) = 2\mu_0 H_c(0)^2/T_c[(T/T_c) - (T/T_c)^3]$, so that $\partial(\Delta S)/\partial T = 2\mu_0 H_c(0)^2/T_c^2[(T/T_c) - 3(T/T_c)^3]$, and $\Delta C = T\partial(\Delta S)/\partial T = 2\mu_0 H_c(0)^2/T_c[(T/T_c) - 3(T/T_c)^2] = -4\mu_0 H_c(0)^2/T_c$ at T_c, with the magnitude therefore equal to $4B_c^2/(\mu_0 T_c)$. Substituting for Nb and Al we then find $\Delta C = 292\,\mathrm{J\,m^{-3}\,K^{-1}}$ in Al, and $\Delta C = 46,100\,\mathrm{J\,m^{-3}\,K^{-1}}$ in Nb.

8.5 The general solution for the magnetic field $B(z)$ inside the plate is $B(z) = A\exp(z/\lambda_L) + B\exp(-z/\lambda_L)$ (eq. (8.26)). As $B(d/2) = B(-d/2) = B_0$, we require $A = B$ and so $B(z) = 2A\cosh(z/\lambda_L)$, with $2A = B_0/\cosh(d/2\lambda_L)$ in order that $B(d/2) = B_0$, as required.

From eq. (8.3), the magnetisation M is given by $\mu_0 M = B - B_0$. For small x, $\cosh x = 1 + \frac{1}{2}x^2$ so that using binomial expansion $B = B_0(1 + \frac{1}{2}z^2/\lambda_L^2 - \frac{1}{2}(d/2)^2/\lambda_L^2)$ and $\mu_0 M = B - B_0 = \frac{1}{2}B_0[z^2 - (d/2)^2]/\lambda_L^2$. Integrating from $-d/2$ to $+d/2$, and then dividing by d, we find $\mu_0 M_{av} = [\frac{1}{3}(d/2)^3 - (d/2)^3]/(d\lambda_L^2) = -d^2/(12\lambda_L^2)B_0$, as required. From eq. (8.36), $G_{sc}(H) - G_{sc}(0) = -\int_0^H \mu_0 M_{av}\,dH = \frac{1}{2}\mu_0(d^2/12\lambda_L^2)H^2 = \frac{1}{2}\mu_0 H_{c0}^2$ at the critical field. We therefore deduce that $H_c(d) = 2\sqrt{3}(\lambda_L/d)H_{c0}$ in the thin film.

8.6 When we integrate eq. (8.62) clockwise around the left-hand side of the loop we find

$$\hbar(\theta_{A1} - \theta_{B1}) = -2e\int_{B1}^{A1} A \cdot dl$$

while for the right-hand side

$$\hbar(\theta_{B2} - \theta_{A2}) = -2e\int_{A2}^{B2} A \cdot dl$$

Adding these together and dividing by $\hbar$ gives $(\theta_{B2} - \theta_{B1}) - (\theta_{A2} - \theta_{A1}) = \Delta\theta_B - \Delta\theta_A = (-2e/\hbar)\oint_C A \cdot dl = (-2e\Phi/\hbar)$, where Φ is the flux linking the loop. This gives $\Delta\theta_A = \Delta\theta_B + 2e\Phi/\hbar$, and $I = I_0\sin\Delta\theta + I_0\sin(\Delta\theta + 2e\Phi/\hbar)$. The total current through the loop then oscillates as Φ is varied; for example, if we set $\Delta\theta = \pi/2$ then it can be shown $I = 2I_0\cos^2(e\Phi/\hbar)$. The period of oscillation is given by $e\delta\Phi/\hbar = \pi$, so $\delta\Phi = h/2e = \Phi_0$, as required. If the two junctions are not identical, the oscillating pattern will be shifted by a half-wavelength, as observed by Wollmann et al. (1993) Phys. Rev. Lett. **71** 2134.

Index